Automated Stream Analysis
for Process Control

VOLUME 1

CONTRIBUTORS

R. NEILL CAREY

A. CONETTA

T. L. C. DE SOUZA

CARL C. GARBER

GERST A. GIBBON

GRETCHEN B. GOCKLEY

JOSEPH P. HACKETT

J. JANSEN

DAN P. MANKA

THEODORE E. MILLER, JR.

R. A. MOWERY, JR.

CRAIG B. RANGER

R. B. ROY

MICHAEL C. SKRIBA

AUTOMATED STREAM ANALYSIS
FOR PROCESS CONTROL

VOLUME 1

EDITED BY

DAN P. MANKA

Pittsburgh, Pennsylvania

1982

ACADEMIC PRESS
A Subsidiary of Harcourt Brace Jovanovich, Publishers

New York   London
Paris   San Diego   San Francisco   São Paulo   Sydney   Tokyo   Toronto

CHEMISTRY

6821-4303

ACADEMIC PRESS, INC.
111 Fifth Avenue, New York, New York 10003

United Kingdom Edition published by
ACADEMIC PRESS, INC. (LONDON) LTD.
24/28 Oval Road, London NW1 7DX

Library of Congress Cataloging in Publication Data
Main entry under title:

Automatic stream analysis for process control.

    Vol. 1: Automated stream analysis for process control
edited by Dan P. Manka.

    Includes bibliographical references and index.
    1. Chemical process control--Automation. I. Manka,
Dan. II. Title: Automated stream analysis for process
control.
TP155.75.A88 1982   660.2'81   82-8822
ISBN 0-12-469001-7 (v. 1)

PRINTED IN THE UNITED STATES OF AMERICA

82 83 84 85   9 8 7 6 5 4 3 2 1

Contents

1 Process Ion Chromatography and Related Techniques

Theodore E. Miller, Jr.

2 Flow-Injection Analysis: A New Approach to Near-Real-Time Process Monitoring

Craig B. Ranger

3 The Monitoring of Cationic Species in a Nuclear Power Plant Using On-Line Atomic Absorption Spectroscopy

Gretchen B. Gockley and Michael C. Skriba

4　The Automation of Laboratory Gas Chromatographs for On-Line Process Monitoring and Analysis

Joseph P. Hackett and Gerst A. Gibbon

5　Process Liquid Chromatography

R. A. Mowery, Jr.

10   Improving the Quality of Infrared Gas Analyzers

Dan P. Manka

11   Waste Gas Analysis Techniques

Dan P. Manka

12   Continuous On-Line Monitoring of Total Organic Carbon in Water and Wastewater

R. B. Roy, J. Jansen, and A. Conetta

List of Contributors

Numbers in parentheses indicate the pages on which the authors' contributions begin.

R. NEILL CAREY (189, 201), Clinical Laboratories, Peninsula General Hospital Medical Center, Salisbury, Maryland 21801

A. CONETTA (303), Technicon Industrial Systems, Tarrytown, New York 10591

T. L. C. DE SOUZA (241), Pulp and Paper Research Institute of Canada, Pointe Claire, Quebec, Canada H9R 3J9

CARL C. GARBER (189, 201), Clinical Laboratories and Department of Pathology and Laboratory Medicine, University of Wisconsin-Madison, Madison, Wisconsin 53792

GERST A. GIBBON (95), Pittsburgh Energy Technology Center, U.S. Department of Energy, Pittsburgh, Pennsylvania 15236

GRETCHEN B. GOCKLEY (69), Westinghouse Research and Development Center, Pittsburgh, Pennsylvania 15235

JOSEPH P. HACKETT (95), Pittsburgh Energy Technology Center, U.S. Department of Energy, Pittsburgh, Pennsylvania 15236

J. JANSEN (303), Technicon Industrial Systems, Tarrytown, New York 10591

DAN P. MANKA (273, 283, 289), Pittsburgh, Pennsylvania 15218

THEODORE E. MILLER, JR. (1), Central Research, Dow Chemical U.S.A., Midland, Michigan 48640

R. A. MOWERY, JR. (119), Applied Automation, Inc., Bartlesville, Oklahoma 74004

CRAIG B. RANGER (39), QuikChem™ Systems Division, Lachat Chemicals, Inc., Mequon, Wisconsin 53092

R. B. ROY (303), Technicon Industrial Systems, Tarrytown, New York 10591

MICHAEL C. SKRIBA (69), Westinghouse Research and Development Center, Pittsburgh, Pennsylvania 15235

Preface

The number of laboratory methods for the analysis of samples has become so numerous that the analytical chemist need only review the literature to find a method suited to his application. Similarly, direct application of analytical methods to continuous analysis of process streams has also grown rapidly. The literature cites many methods for analysis, such as gas and liquid chromatography, atomic absorption, infrared, and ion chromatography, giving the chemist a choice of instrument to use.

In many industries the chemist retrieves snap samples from the process stream. The analysis of these samples comes too late for the operator or chemical engineer to make changes in the control of the process. Besides the advantage that continuous analysis has in this respect, there are many others, such as an increased yield of a pure chemical in a distillation process and a reduction in variation of quality of manufactured products in the petroleum and chemical industry. Also to be considered are increased plant capacity, reduced deviations (thus making the product more uniform), and reduced energy consumption and operating costs. There is less reprocessing of off-specification material, which leads to better matching of customer specifications and ultimately to increased sales. In the petroleum industry, process analysis contributes to longer catalyst life and a reduction in coking and distillation flooding. In representative distillation column applications, chromatograph-based control has been found to reduce energy requirements by 10–20%. Other benefits include a 5–10% throughput increase, a 5–30% value increase in the product, and smoother operation of the fractionator.

The potential returns from optimizing catalytic crackers are high because units not only incorporate feed heaters that consume significant quantities of fuel gas but also are large producers and users of steam. For example, in one two-unit complex, an analyzer-based information system was critical to implementing changes that reduced gas consumption by about 100 million BTU per hour, generating a return of over $1 million annually.

Before a chemist or chemical engineer decides to take advantage of continuous analysis, he should spend considerable time in the plant to determine all the ramifications of the process. Only then should he seek the specific method of analysis that would benefit the complexity of his particular stream. There is much more to analyzing a process stream than finding a suitable method. It is the intent of this book to put together the experience of a number of experts who have successfully developed stream analyzers, detailed the entire system, specified special controls, supervised its construction, and brought it to the point of being operational. Other elements to be considered, besides selection of the analyzer, are the location of the sampling probe, the construction materials for systems that will be located in a harsh environment, and the filtering of the sample by various methods to clean it before injection into the analyzer. Then the results of the analysis must be used properly by the computer in order to adjust the process controls so that the process stream will reach its ultimate goal.

These are the subjects that will be covered in this book by experts who have successfully followed the outline above to reach their goal. Even though control of the exact process by the chemist, chemical engineer, laboratory manager, or plant manager is not listed in the chapters, it is hoped that the reader will learn so much from the descriptions of the newest developments in the atomic energy, chemical, coal gasification and liquefaction, pulp and paper, and steel processes that he will feel that he has received some guidance in solving problems that could be encountered in his own future installation.

If a chemist or chemical engineer decides to develop a process control method, he will normally put much time and effort into sample points, sample preparation, and sample transport with many negative results before attaining a satisfactory method. However, with this book he can move forward rapidly. He can choose or adapt, with some modification, procedures that are the result of many tests by the authors. These procedures have been used successfully in various applications.

In this volume, a group of experts have prepared chapters on key aspects of automated stream analysis. Specific subjects treated include process ion chromatography, flow-injection analysis, on-line atomic absorption spectroscopy, on-line gas chromatography, and process liquid chromatography. Infrared gas analyzers are also covered and a series of essays provide discussions of stream analyses in various industries.

It should be possible for the individual working in any of these areas to obtain a good review of the individual's specific area and an overview of related areas.

We trust that we have been able to bring the reader an up-to-date survey of the major areas of research and development and to convey a number of the outstanding problems that remain to be solved together with some of the approaches that may be used toward that end.

1

Process Ion Chromatography and Related Techniques

THEODORE E. MILLER, JR.

Central Research
Dow Chemical U.S.A.
Midland, Michigan

I. Introduction

Complex analyzers play a significant role in the modern chemical production plant. For example, among the varied and growing assortment of instrumental methods available for totally automated process analysis, a technique whose use is particularly widespread is gas chromatography (GC). Why should this hold true for an analytical procedure that calls for greater complexity than the in situ process probe? The answer is that in many cases the application itself demands measurement of a trace impurity in a diverse mixture or complete characterization of the entire mixture. In such instances the powerful specificity of GC justifies its selection, regardless of the additional special care needed in implementation.

As improvements are made in liquid minipumps, valves, and associated hardware, there is a movement to exploit liquid chromatography on-stream more fully along with a broad range of associated nonchromatographic flow-injection techniques.

This chapter illustrates some uses of quite recently developed liquid flow-injection analytical devices in process control applications within the chemical production plant. The initial two techniques, ion chromatography and ion exclusion chromatography, are currently used to analyze over 90 varieties of ions down to parts-per-billion in aqueous streams (Maugh, 1980) in hundreds of documented laboratory applications in government, academic, and industrial laboratories. We shall describe here totally automated process versions in the chemical industry.

The final two approaches treated in this chapter have as yet had relatively little laboratory and process use but are included nonetheless since they exemplify the rather remarkable variety of analytical/process control capabilities of similar hardware components in different arrangements.

A special emphasis in the beginning is accorded to sample handling practices, an absolutely vital aspect of flow-injection process instrumentation.

II. Sample Preparation for Process Analyzers

A. Basic Guidelines

There are certain requirements of any on-stream analyzer liquid sample conditioning system that are quite essential:

(a) the process sample introduced to the instrument must be physically compatible with it in terms of pressure, abrasiveness, and corrosivity in order to avoid mechanical damage;

(b)   the sample introduction ought to be rapid enough to satisfy process control response time conditions;

(c)   the sample as presented in the analyzer is required to be truly representative of its actual composition in the plant process; and

(d)   provisions for *safe* maintenance are mandatory.

Physical compatibility between the sample and analyzer is typically achieved through the use of components such as pressure regulators, atmospheric break devices, filters, diluters, and extraction systems along with proper selection of materials of construction. The seemingly excessive costs of exotic, corrosion-resistant materials of construction for process sampling systems can easily be dwarfed by the maintenance and down-time expenses incurred without them. Since experience has shown that the majority of process analyzer down-time stems from sample system failures, a cost of $20,000 for a sampling system is clearly not excessive for a $50,000 analyzer in a high-payout application.

The requirement for sufficiently rapid sample delivery necessitates judicious downsizing of sample lines, and a means of continuous flow monitoring in the system is to be preferred. Any provisions for test standard admission to the instrument are best located so as to include as much of the sampling system as practicable in order that inherent delays can be discovered and remedied.

Ensuring that the delivered sample is representative of the process stream itself begins at the process take-off. A sampling probe inserted to center-pipe tends to avoid deleterious stagnating wall effects, settled particles, and trapped gases. Again, rapid flow, large diameters, and inert materials all serve to minimize selective surface adsorption or degradation of key sample components along the way to the analyzer. Ultimate assurance of representative automated sampling can come from regular manual sampling and analysis from altogether separate process take-off points.

Safe operation of sample handling systems is essential in the chemical plant, where analyses are routinely performed on corrosive acids or bases, toxic wastes, and other hazardous substances. Safety is aided by means of duplicate pressure gauges, outlets to atmospheric pressure rather than to the process, easy purging through valve operation, interlocks to guarantee safe line-preparation sequences, and the elimination of frangible observation barriers.

B.   Direct Approaches

''The simpler, the better'' is an adage that surely applies to process analyzer sampling systems. Based on industrial experience with hundreds of fully automated on-line instruments, certain key elements of the application rather accurately predict the relative likelihood of success for a proposed analyzer. These are listed in order of impact in Table I.

TABLE I

KEY PREDICTORS OF PROCESS ANALYZER SUCCESS

Reliability detractors	Reliability enhancers
1. Mechanical analyzer components	No moving parts
2. Special sample manipulation (extraction, dilution, phase changing, etc.)	Direct sample analysis
3. High sensitivity requirement	Low sensitivity requirement
4. Toxic and/or flammable sample	Safe sample
5. Particulates in sample	Solids-free sample
6. One-of-a-kind prototype	Well-established method
7. Corrosive sample	Noncorrosive sample
8. Sample multiplexing	Single sample
9. Liquid or solid sample	Gas sample
10. Rugged ambient environment	Mild environment

It is significant that six of these ten criteria are sampling-related and that several of the "reliability detractors" in Table I result from special sample manipulation rather than direct analysis.

For this reason the chapter begins with an outline of relatively simple, direct sample conditioning approaches and later progresses to examples of more exotic manipulative methods recommended only when direct alternatives have proved inadequate.

Bypass filtration, illustrated in Fig. 1, is generally preferred over flow-through filtration. As portrayed in the figure, process stream movement tends to sweep potential accumulations from the filter element while conveying heavier particles in suspension away from the filter's porous surface. It is necessary, however, to maintain back-pressure at the filter outlet to drive the split filtered stream through the porous filter matrix and on to the analyzer. Another advantage of the bypass arrangement is greater flow from the process to the filter inlet and reduced sample lag time. Increased differential pressure readings on optional gauges installed at the inlet and analyzer outlet ports of either filter can provide a convenient and safe indication of element plugging.

Depending upon the analysis technique, overfiltration can be as serious a source of trouble as insufficient filtration since frequent filter plugging and down-time are the probable outcomes.

Testing filtration alone on an actual process sample prior to the arrival at the plant of the instrument can expedite a smooth startup.

A system for returning the sample directly to the process is shown in Fig. 2. This scheme allows for easy verification of flow and safe analyzer leg purging for maintenance purposes. Flow status is inferred from the readings on gauges 1 (G1) and 2 (G2). Flow through the analyzer produces a pressure drop across

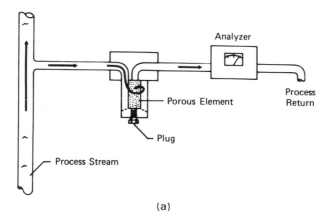

(a)

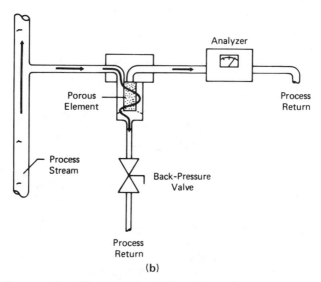

(b)

Fig. 1. Process analyzer filtration modes: (a) flow-through filtration, (b) bypass filtration.

throttling valve C and a differential G1, G2 reading. To eliminate the possibility that valve C is plugged, closing either valve A or E will rapidly alter both gauge readings if valve C is not plugged. An identical reading on G1 and G2 thus directly indicates insufficient flow at the analyzer.

As in Fig. 2, purge fluids and gases may be admitted by closing valves A and E and diverting valves B and D. For increased instrument isolation from the process during maintenance the "double-block-and-bleed" arrangement may be

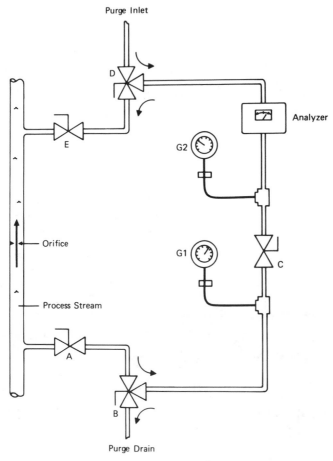

Fig. 2. Process return sampling system.

used, which, in place of a single block valve A or E, substitutes a vent valve interposed between a pair of block valves. Leakage through such a system requires the unlikely simultaneous failure of three valves.

C. *Continuous Partial Distillation*

Occasionally it is necessary to perform automated on-line analysis of volatile compounds in aqueous brine streams. Trace contaminants such as phenols, substituted phenols, aromatics, and alcohols can be removed in plant brine purification units, enabling the recycling of these raw brines and thus avoiding expensive disposal. However, process monitoring is needed to ensure that clean-up units are operating within specifications.

These samples may present formidable process analysis challenges when, for example, parts-per-million alcohols in 20% salt brines are the analysis objective. Such corrosive high-chloride solutions may contain high particulate loads and leachates from process piping. Unfortunately, rugged in situ analytical probes are overwhelmed by the abundance and variability of interferences and defy calibration. For many of these applications, flow-injection methods can provide the needed measurement specificity, but, because of the injection valve and

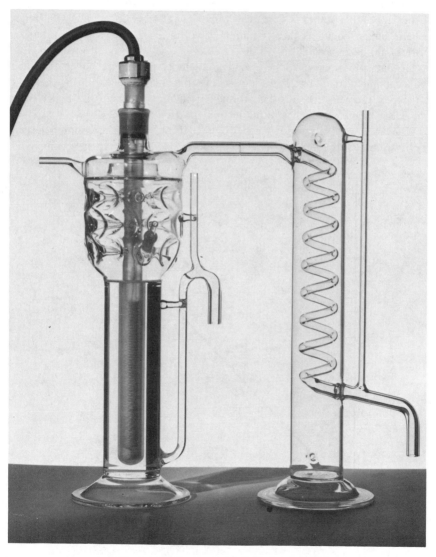

Fig. 3. Continuous partial distillation apparatus.

critical internal orifices, flow-injection instrumentation demands thorough sample cleanup.

Continuous partial distillation can be a useful sample conditioning technique for monitoring volatiles in aqueous streams in an uninterrupted, automated fashion. One embodiment of this approach appears in Fig. 3. This system admits process sample from an overflowing constant-head tank to the upper left port of the "boiler," a 12-in. glass unit containing a 600-W quartz immersion heater. A distributor ring and internal projections maximize steam vapor contact with the incoming sample stream. As shown in the figure, an overflow leg with atmospheric break maintains the "bottoms" level in the boiler. No fine filters or orifices are necessary.

The distillate proceeds to the "condenser" to the right of the boiler. Provided with cooling water through the ports projecting rearward in Fig. 3, this unit efficiently recovers the distillate formed from the sample in the boiler. The condenser effluent proceeds to the instrument itself.

What about recovery ratios for the compounds of interest, information needed for analyzer calibration? A series of experiments comparing key compound

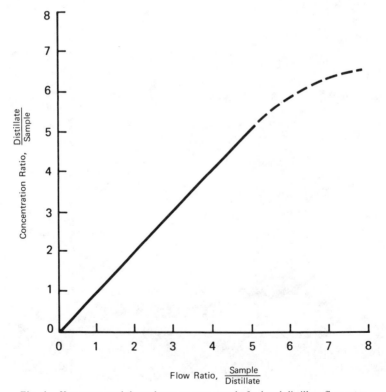

Fig. 4. Key compound dependence on raw sample feed and distillate flow rates.

levels in raw sample feed, boiler overflow and distillate may be used to characterize the calibration. For the example of alcohol recovery from concentrated
brine shown in Figure 4, the desirable result is that the sample concentration
ratio, distillate-to-feed, is equivalent to the feed-to-distillate flow rate ratio, the
outcome expected for 100% efficient capture in distillate of the trace alcohol
originally present in feed. The significance of this direct relationship is that

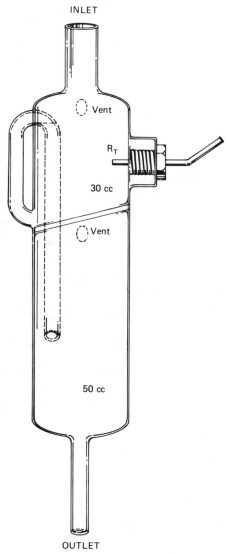

Fig. 5. Siphon dump assembly.

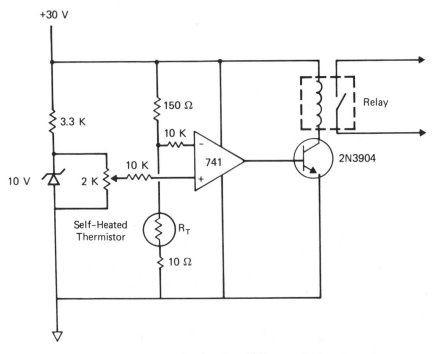

Fig. 6. Self-heating thermistor fluid sensor circuit.

merely measuring feed and distillate flow rates furnishes the instantaneous sampling system concentration factor.

One alternative method of measuring flow rates in this range at atmospheric pressure is the siphon dump assembly, illustrated in Fig. 5. When the upper vented compartment fills to the top bend of the overflow tube, a sustained siphon effect suddenly empties this chamber into a vented lower compartment. The siphon is broken after the fluid batch has fallen, and another cycle begins. Level sensor R_T, a self-heated thermistor arranged as in Fig. 6 to detect the greater thermal dissipation of the fluid, indicates to a timing system or computer the dump events, and, ultimately, the flow rate is computed. On this basis, instantaneous corrections in concentration ratio can be performed.

For analyzing alcohols and highly volatile light hydrocarbons in aqueous brine streams, continuous partial distillation offers

(a) mechanical simplicity, no moving parts;
(b) particulate-free distillate sample;
(c) brine removal, reduced analyzer corrosion;
(d) measurable key compound concentration factor.

D. *Dilution*

A number of liquid process samples in the chemical industry unavoidably require volumetric dilution into some intermediate solvent prior to introduction into the analyzer. Such samples often include high-melting-point organic substances as pictured at room temperature in Fig. 7. The rock-like material shown

Fig. 7. Organic sample with analysis column. [Reprinted with permission of the Instrument Society of America, 1980, from articles appearing in *Adv. in Instr.* **33–35.**]

is an intermediate in the production of a herbicide. Tightened process control, based upon on-line determination of the individual isomers comprising this molten sample, increases the manufactured-product production capacity of a single plant by an amount well over $1,000,000 per year.

But one might wonder why dilution should be attempted, with its attendant complexity, in an entirely automated on-stream analytical application. First, analytical methodologies with adequate specificity for separation and measurement of nearly identical, isomeric compounds require that extremely small amounts of sample mass be presented to the analyzer (Snyder and Kirkland, 1979). The tiny vertical glass tube in Fig. 7 illustrates the comparative scale of this sample and a typical flow-through liquid chromatographic column. The challenge is to condition a minute fraction of the rock-like sample for admission to the miniature tube for analysis. Also, dilution into the solvent dramatically lowers the freezing point of the mixture, eliminating the likelihood of wax-up and plugging of the instrument lines at some inadequately heat-traced spot. A further benefit of dilution is that particulate matter is reduced in concentration along with the sample itself, and subsequent filtration becomes less necessary. Finally, safety is enhanced since the analyzer contains predominately a cooler, less-toxic diluted sample at lower pressures.

A recently developed dilution system that has operated quite satisfactorily for on-line analysis in the chemical process industry (Miller and Cabala, 1979) was designed to offer the following features:

(a)   few mechanical components, e.g., pumps, valves likely to require maintenance;
(b)   no capillaries or orifices to plug or vary;
(c)   conservation of diluent;
(d)   minimum delivery time and hang-up;
(e)   inherent safety via all-pneumatic operation; and
(f)   ±1% repeatability for analytical precision.

This diluter apparatus is pictured in Fig. 8, and a corresponding operating schematic appears in Fig. 9. Referring to the latter figure, there are four principal steps in its operating sequence. (i) *Fill:* The six-port air-operated plastic valve to the central lower portion of the figure actuates, permitting the applied 30 psi air pressure to force the diluent initially contained in the external loop onward to the process valve, shown in detail in Fig. 10. Thus, a precisely fixed volume of diluent is conveyed to the chamber above inlet F, where it remains in bubbling agitation. Enlargement of the port C diameter prevents plug flow upward beyond the process valve. Next (ii) *Inject:* The air-actuated piston D withdraws a volume of pure sample as established by the dimensions of the notch E from the left-hand process region to the waiting, agitated diluent in the right-hand valve portion.

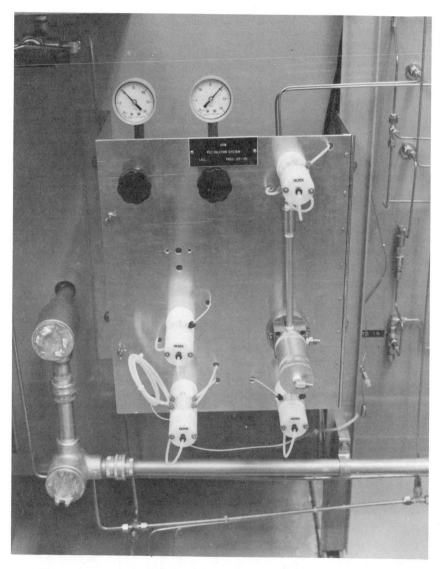

Fig. 8. Process sample dilution apparatus. [Reprinted with permission of the Instrument Society of America, 1980, from articles appearing in *Adv. in Instr.* **33–35.**]

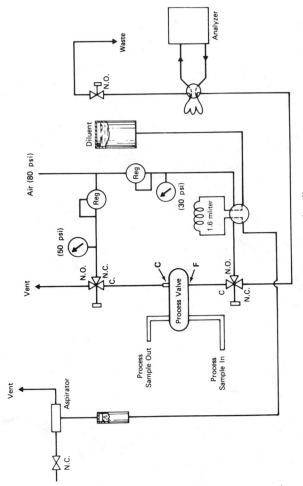

Fig. 9 Diluter operation diagram.

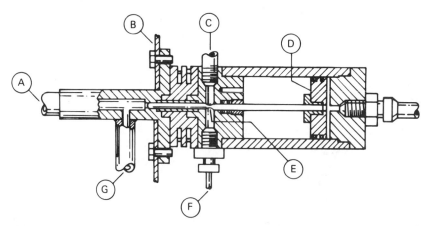

Fig. 10. Dilution system process injection valve. A, process sample outlet; B, mounting panel; C, dilution chamber vent; D, air-actuated piston; E, injection notch; F, diluent inlet; G, process sample inlet.

Although a temperature gradient is maintained across this valve assembly, with the heated zone lying to the process side and a cooler region on the actuator side, solidification of the sample on the stem is prevented by its immediate immersion into the agitated, bubbling diluent.

The next stage is (iii) *Feed:* The pair of three-way valves above and below the process valve, shown in Fig. 9, forces diluted, mixed sample at 50 psi on to a collecting vessel or analyzer injection valve. By what means can a rather small aliquot of the diluted sample be delivered under pressure to an injection loop with maintenance-free, nonmechanical components and yet be stopped in this position to await indefinitely the analyzer injection event? A block valve downstream of the analyzer injector closes prior to the feed step in the dilution sequence so that the delivery pressure works plug-flowing diluent into a trapped volume of air beyond it. Surface tension and small tubing diameters assure plug flow. Appropriate tubing volume is estimated from Boyle's Law (Daniels and Alberty, 1966), assuming isothermal ideal gas conditions,

$$P_i V_i = P_f V_f \tag{1}$$

and so

$$\frac{V_f}{V_i} = \frac{P_i}{P_f} \cong \frac{15 \text{ psia}}{(50 + 15) \text{ psia}} = 23\%, \tag{2}$$

where P_i and P_f are atmospheric pressures at initial (ambient atmospheric) and final (50 psig pressure) conditions and V_i and V_f are the corresponding gas volumes. Having established the tubing diameter and the volume reduction under pressure, it is a straightforward matter to locate the analytical injector close to the center of the delivered diluent volume by the choice of tubing lengths. Exact positioning is not critical, however, since the sample is uniformly mixed with diluent.

One step remains to complete the dilution cycle. (iv) *Load:* The external loop diluent-metering valve remains connected at this point to balanced hydrostatic columns of diluent. The metering external loop contains air held in place by these equal pressures. To eject the residual air, the air-powered aspirator of Fig. 9 is briefly activated, sucking diluent forward through the loop from the diluent

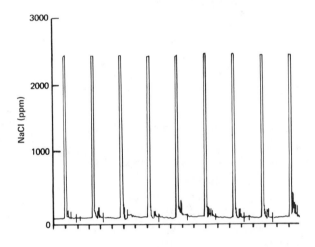

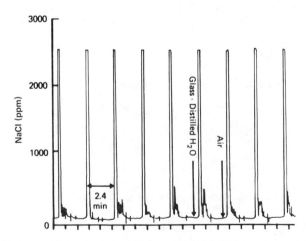

Fig. 11 Dilution precision and mixing uniformity. [Reprinted with permission of the Instrument Society of America, 1980, from articles appearing in *Adv. in Instr.* **33–35.**]

reservoir. The undesirable bubble is simply vented through the aspirator when it reaches the vertical piping. The diluent level in the vertical leg rises only slightly during the allotted time, however, owing to enlarged diameters. When the aspirator is shut off, diluent levels reequilibrate. The net result is that absolutely no diluent is wasted.

The peaks in Fig. 11 show the precision and uniformity of dilution. In this test the system diluted a 10% sodium chloride solution 40:1 in deionized water and used a flow-through conductivity cell in place of the analyzer injection valve. It is calculated that the dilution factor repeatability here and in extended process applications is easily within 1% at the 95% confidence level. The flatness of the peaks in Fig. 11 exemplifies the excellent uniformity of mixing achieved by the simple bubbling technique.

Owing to the plug-flow line purging during each dilution cycle, there is no measurable previous sample hang-up problem.

Diluter maintenance time is on the average less than 2 hours per month in continuous on-stream applications.

Besides its use in the preparation of high-boiling organic samples for process analysis by dissolving them in diluting solvent, a modified version of this system is also used to extract additives from insoluble solids for flow-injection analysis (Bell and Miller, 1980). The process valve is replaced by a disposable cartridge manually packed with the insoluble solids of interest. The extractant, again delivered pneumatically, agitates the sample solid and is subsequently forced through a filtering surface to separate residual solids from the extractant solution of analytical interest. Otherwise, the methods are quite similar.

III. Process Ion Chromatography in Power Production*

A. *Objective*

On-line monitoring for trace corrosive ion species in power plant waters is becoming more and more necessary owing to increases in demand for electrical energy. Thirty years ago, 200-MW power plants were among the largest. Today, however, 1000-MW plants are typical, and nuclear plants are in the 1300-MW range. Because of the immense capital investment in these more recent plants, careful continuous monitoring for trace corrosives is imperative.

Ion chromatography is used to detect anions in power plant steam samples at Westinghouse (Borman, 1980). Chloride contamination that is as low as parts per billion has a proven corrosive effect on turbine blades. Monitoring these trace anions requires that steam samples be taken manually to a laboratory ion chromatograph.

*See also Chapter 3.

At The Dow Chemical Company, fully automated on-line ion chromatographs measure anions in power plant boiler water (Stevens *et al.*, 1977). Here, the object of protection against corrosion is the internal surface of the boiler itself. Boiler tube corrosion and scale buildup are, in fact, reduced by blending parts-per-million chemical additives into boiler water. Two key additive compounds are orthophosphate and sodium sulfite. Orthophosphate, maintained between 10–20 ppm as Na_2HPA_4, inhibits the scale-forming divalent cations calcium and magnesium while also coating and passivating tube surfaces. Sodium sulfite, generally held from 5–10 ppm as Na_2SO_3, removes excess oxygen from boiler water via sulfate formation and thereby suppresses pitting corrosion within the boiler tubes. Continuous analysis provides assurance of sufficient additive blending into boiler water.

The key feature of ion chromatography is its so-called "eluant suppression" step.

B. Eluant Suppression

Although the technique of ion chromatography is only seven years old (Small *et al.*, 1975; Small and Solc, 1976), its use is already widespread in analyzing more than 100 varieties of ions, organic and inorganic, in process fluids, wastewaters, food products, auto exhaust, and air samples (Mulik and Sawicki, 1979). Its appeal derives from its high sensitivity and selectivity combined with wide versatility.

Ion chromatography is an ion exchange chromatographic method wherein the stationary phase bears ion exchange functional groups. A cation exchanger, for example, is preferably composed of a pellicular low-capacity sulfonated polymeric resin (Stevens and Small, 1978). The anionic sulfonic group is covalently affixed to the relatively inert plastic resin support so that cations associate with the resin only by electrostatic attraction and can be easily displaced by other cations. The sulfonate groups are maintained in an ionic form determined by the flowing eluant and retain cations according to the combined effects of the ion affinity sequence and relative ion concentrations. When a column packed with this sulfonated resin, initially containing a mixture of cations, is eluted with dilute sodium chloride electrolyte, for example, more weakly bound cations are eluted first, most strongly bound cations emerge last, and the others appear in the intermediate elution volume.

The principal innovation in ion chromatography resides in its method of ion detection in chromatographic separator column effluent. Note that in the sodium chloride elution of several ions from a column, as described above, the sodium ion is continually present in the electrolyte background, overwhelming any distinctive signal arising from other cations. "Eluant suppression" removes this obstacle to detection.

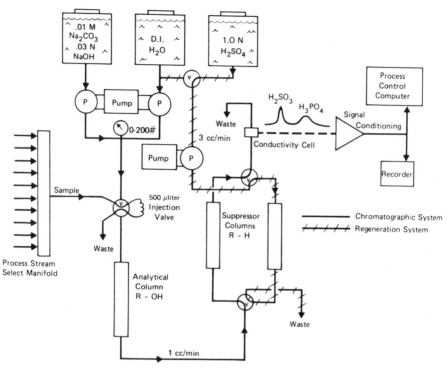

Fig. 12 Process ion chromatograph for boiler water analysis. [Reprinted with the permission of the Instrument Society of America, 1980, from articles appearing in *Adv. in Instr.* **33–35.**]

The suppressor column, as shown in the process analyzer schematic, Fig. 12, is filled with an ion exchange material of an ion exchange polarity opposite to that of the separator column and immediately follows the separator. It continually converts relatively highly electrically conductive eluant solutions to a low conductivity form.

Referring to Fig. 12 and considering anion analysis, we see that the separator is packed with an anion exchanger which is most often manufactured by amination of polymeric colloidal particles, which are agglomerated with cation exchange support resin beads to form fixed amine cation sites (Small and Stevens, 1978). Hydroxide ions sequentially elute the sulfite and phosphate ions as peaks from the separator column, but without further reaction, conductimetric detection would be overwhelmed by the hydroxide ions of eluant. Simply diluting the hydroxide to bypass this detection difficulty either sacrifices resolution of the sulfite and phosphate peaks from each other or unacceptably prolongs elution time.

The cation-exchanging suppressor column is maintained in the hydrogen ion form (R–H$^{\oplus}$) so that the ion exchange processes therein are as shown below.

Eluant	NaOH	R–H$^{\oplus}$	H$_2$O
	Na$_2$CO$_3$	$\xrightarrow{\hspace{2cm}}$	H$_2$CO$_3$
Sample	Na$_2$SO$_3$	Suppressor	H$_2$SO$_3$
	Na$_2$HPO$_4$	column	H$_3$PO$_4$

(3)

Species formed from the eluant, i.e., water and carbonic acid, are low in conductance whereas the sample species are converted to their conjugate acids, highly conductive forms. The result is excellent sample ion conductivity against a background of low-conductance. A typical ion chromatographic output is shown in Fig. 13.

The limits on the size of the suppressor column are dictated on the one hand by the need for sufficient capacity to accomplish a complete exchange of the eluant during an analytical run and on the other hand by the usual chromatographic requirement to minimize volume between the separating column and detector.

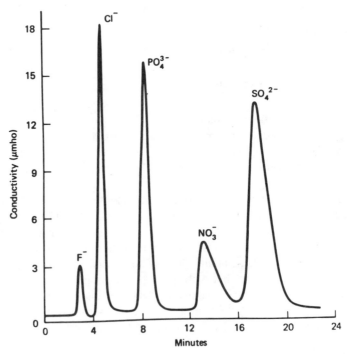

Fig. 13 Example of an ion chromatogram.

Too great an interposed volume produces deleterious peak spreading and loss of chromatographic efficiency.

In the automated, on-stream ion chromatograph, dual suppressor columns are automatically interchanged with each sample injection.

C. Design Features

The essential features of the boiler water ion chromatograph for sulfite and phosphate appear in Fig. 12. Photographs of these instruments have been published elsewhere (Miller, 1978). This system automatically sequences the sample stream selection via a commercial process GC digital programmer. The sample is bypass-filtered prior to arrival at the stream select manifold.

Stream valve switching occurs immediately after injection of the prior stream to allow maximum purge time through the sample injection valve of the ion chromatograph itself. An air-actuated plastic valve designed for liquid chromatographic use is preferred. Sample ion quantitation is based upon peak size, and therefore any irreproducibility of injection volume is undesirable.

Pump and associated column hardware items are again available commercially from liquid chromatography component suppliers. With the eluant filtered to 1 μm before introduction to the pump, more than one year of continuous, stable pump operation is not unusual.

The conductivity detector cell must be miniaturized to be compatible with the scale of the other chromatographic components. A conductivity metering circuit that is quite suitable for commercial miniature cells, and which is most often employed in process ion chromatographs at The Dow Chemical Company, is outlined in Fig. 14.

The relationship between ion concentration and specific conductance for sodium chloride is plotted in Fig. 15. The temperature compensation designed into the circuit in Fig. 14 corrects for this inherent temperature dependence. Note also the linear relationship between conductivity and concentration at these relatively dilute levels of ions in water.

The conductivity meter output feeds into the standard chromatograph detector input channel of a commercial process gas chromatograph programmer. This processor performs baseline drift corrections and continually outputs ion concentration data based on the most recent set of peak heights.

Alarms are generated if ion concentration output signals fall above or below preset levels, and the boiler water chemical additive problem can be immediately corrected.

Miscellaneous automatic warning signals alert plant personnel to incipient problems involving eluant reservoir level, sample pressure, pump pressures, and electrical baseline levels.

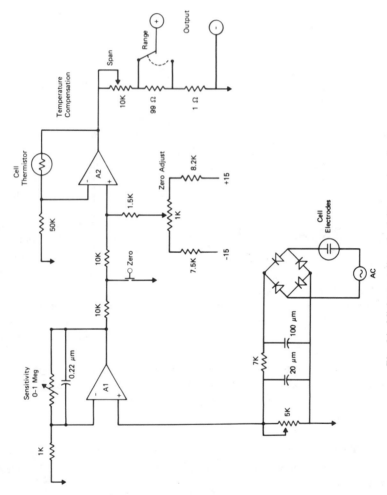

Fig. 14 Miniature cell conductivity metering circuit.

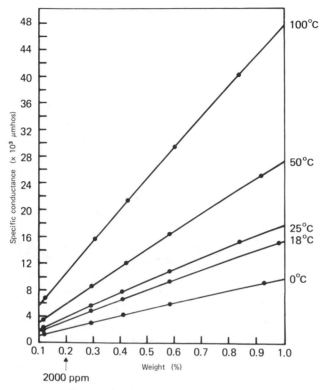

Fig. 15 Sodium chloride concentration related to electrolyte conductivity. Measured conductivity = specific conductance/cell constant. [Reprinted with the permission of the Instrument Society of America, 1980, from articles appearing in *Adv. in Instr.* **33–35.**]

D. Results

Successful performance of the power plant ion chromatographs has been achieved only by recognizing certain crucial requirements:

(a) routinely, at least daily, it is necessary to survey the instrument for excessive pressure or leaks and to correct these problems early;

(b) separating column effective lifetime can be extended beyond one month by utilizing high purity water for eluant;

(c) sample line filter cartridges must be replaced frequently; and

(d) plant personnel need to budget up to eight man-hours per week for routine maintenance of each system.

The results that justify this attention, however, are frequent updates of valid, recorded information on critical corrosion-suppressing additives rather than, as in the past, questionable data from wet chemical analytical methods infrequently carried out by plant personnel with unpredictable analytical skills.

IV.   Ion Exclusion Chromatography for Brine Purity

A. *Purpose*

Not long ago, deep well disposal was the fate of waste brine solutions from chemical production facilities. Recently, however, aqueous waste streams containing more than 10% salt have been recognized as worthwhile raw materials to be recycled into production, provided contaminants accumulated from chemical processing can be removed or held within specifications.

Another form of flow-injection analysis, ion exclusion chromatography (IEC), has been incorporated into on-line instrumentation to ensure that particular trace ionic contaminants fall within acceptable concentration limits in process waste brines. Such monitoring allows the recycling of brine in lieu of energy-intensive shipping alternatives or deep-well disposal.

Specifically, ion exclusion chromatography is adapted to the automated, on-line analysis of 0–200 ppm by weight sulfuric, acetic, and propionic acids in 20% sodium chloride brine in the example selected here.

B. *Ion Exclusion Separation*

The method for ion exclusion separation of these acids shown in Fig. 16 is based on chromatographic injection of the acid mixture onto a cation exchange resin (Richards, 1975), unlike the ion chromatographic technique above wherein various anion species separate in anion exchange columns. Ion chromatography is abandoned here because of the advantage of using a simple deionized water eluant in an automated instrument besides that of the absence of eluant suppression and associated regeneration hardware, both simplifications owing to the nature of ion exclusion chromatography. Unfortunately, IEC applies solely to separation and analysis of compounds that have distinct dissociation constants, and it cannot begin to approach in scope the wide applicability of ion chromatography.

How does an acidic cation exchange resin-packed column perform chromatographic separation of a mixture of acids using a water eluant? Relative acid strength in aqueous solution is usually expressed in terms of the equilibrium ionization constant K such that

$$HA \overset{K}{\rightleftharpoons} H^+ + A^-, \tag{4}$$

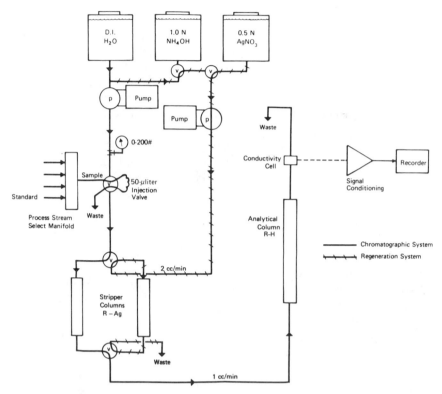

Fig. 16 Schematic of ion exclusion chromatograph with brine stripping for mixed acid analysis. [Reprinted with the permission of the Instrument Society of America, 1980, from articles appearing in *Adv. in Instr.* **33–35.**]

where

$$K = [H^+][A^-]/[HA] \tag{5}$$

and brackets denote quantities in moles/liter. Since by definition

$$pH = -\log [H^+] \quad \text{and} \quad pK = -\log K, \tag{6}$$

substitution and rearrangement of Eq. (5) yields the relationship

$$pH = pK + \log [A^-]/[HA]. \tag{7}$$

From this expression, the extent of acid dissociation can be determined by comparing the characteristic constant pK for the acid under consideration with the pH of its solution. Note that the pK is equal to the pH at which the logarithm term in Eq. (7) vanishes, that is $[A^-]/[HA] = 1$, and therefore at this particular pH equal amounts of acid are in the ionized and nonionized forms. Raising the pH (increasing alkalinity) augments the proportion of ionized acid, whereas lowering it drives a greater fraction of moderate pK acids into their nonionized forms.

By employing an acidic ion exchange separation resin with a relatively hydrophobic matrix, the ion exclusion column suppresses the ionization of the acids in the sample differently and thus elutes them in a sequence of increasing pK with a simple water eluant. Weaker acids (of higher pK) exhibit relatively suppressed ionization and thus more strongly partition into the stationary phase, which partly accounts for their later appearance in the elution sequence.

Separation is also aided by differences in sample-resin hydrophobic interactions among nonionic acids of varying hydrocarbon chain length.

Detection of the separated sample species is the final chromatographic step.

C. Brine Stripping

As in ion chromatography, conductivity detection is an appealing method of detection in process instrumentation since it offers

(a)   miniature internal volume,
(b)   no routine maintenance,
(c)   sizable signal-to-noise ratio, and
(d)   universal sensitivity to ionic species.

A problem arises when we perform the concentrated sodium chloride brine analysis attempted here, however. To detect weak acid contaminants at 20 ppm in 20% brine, the method must detect one part sample in an overwhelming 10,000 parts chloride. An ion exclusion column, in spite of excellent efficiency, allows fast-eluting chloride to tail into and swamp the appearance of the critical sulfate, acetate, and propionate peaks.

To remove most of the chloride before the injected sample encounters the ion exclusion separator, a "brine stripping" column (Small and Stevens, 1975b) is packed with another high-capacity cation exchange resin in the silver ion form and arranged in the system as shown in Fig. 16. The stripper column exchanges sample cations, mainly sodium, for silver ions, which immediately form a silver chloride precipitate that remains behind in the stripper column. In the stripper overall,

$$
\begin{array}{lcl}
& \text{R-Ag}^{\oplus} & \\
\text{Na}_2\text{SO}_4 & \Rightarrow & \text{Ag}_2\text{SO}_4 \\
\text{Na Acetate} & & \text{Ag Acetate} \\
\text{Na Propionate} & & \text{Ag Propionate} \\
\text{NaCl} & & \text{AgCl} \downarrow
\end{array}
\qquad (8)
$$

The acid anions pass through the stripper and onto the ion exclusion column, where they separate.

As is usual in chromatography, too great a volume interposed between the injection valve and the analytical column produces undesirable peak spreading, so that the stripper cartridge must be as tiny as possible. And yet, the ion exchange capacity of the stripper ought to be sufficient to neutralize completely the sample chloride load. Both requirements are met by an automated stripper interchange routine that frequently regenerates one tiny column while the other is actively engaged in stripping brine from sample.

In the regeneration phase, the silver chloride precipitate is flushed from the spent stripper by means of an ammonium hydroxide solution that quite rapidly forms a soluble silver ammonium complex. A silver nitrate rinse follows, restoring the ion exchange resin to the silver form for the subsequent sample stripping process.

The number of injections a regenerated stripper can accommodate is expressed as

$$N = \frac{C \times V}{S \times v},$$

(9)

where N is the number of sample injections, v the injection volume (cc), S the sample concentration (meq/cc), C the resin specific capacity (meq/cc), and V the stripper column volume (cc).

There is no appreciable loss in the ion exchange or brine removal capacity in spite of hundreds of stripping cycles in the automated instrument.

D. The On-Stream Analyzer

The process ion exclusion chromatograph is fabricated from components identical to those of process ion chromatographs, with real differences only in the composition of eluants and the placement of ion exchange columns. As is evident in Fig. 16, a conductivity cell again generates a chromatographic peak pattern to further signal-conditioning electronics or perhaps only a local chart recorder from which operators can interpret sample consituent concentrations from peak height.

Note that certain sequential events must be performed by the analyzer, i.e.,

(a) stream selection,
(b) sample injection,
(c) stripper column alternation, and
(d) sequencing three regeneration solutions to the off-line spent stripper column.

As portrayed in Fig. 16, chromatographic valve switching accomplishes each of the above steps by means of pneumatic actuation through pilot solenoid valves. Air actuation permits isolation of electrical devices from chemical solution-handling elements of the analyzer, as is desirable.

Once during each cycle, the process stream-select manifold admits an internally stored standard sample containing a representative and accurately known mixture of acids in brine. A periodic identical chromatographic pattern from standard solution assures plant operators that the instrument is operating satisfactorily.

An indispensable feature of the analyzer is a control panel with override switches and lamps to display the status of analysis: stream identification, injection, stripper column mode, and regeneration phase.

E. Maintenance

An instrument of this type operating continuously demands daily cursory inspection of pressure readings and bypass flow of sample and eluant to waste. The slightest leaks of corrosive brine must be corrected immediately. Over a longer term, the pumps require oiling as well as general servicing; also the ion exchange resins gradually lose exchange capacity over extended months' use and must be replaced. Careful 1-μm filtering of the sample and use of a deionized water eluant significantly prolong the effective lifetimes of the pumps and ion exchange column packings.

Overall, ion exclusion chromatographs in continuous on-stream applications can demand as much as eight hours of maintenance by plant personnel every week. But even this level of maintenance is a vast improvement over the time-consuming, frequent bench analyses carried out previously.

V. Total Dissolved Solids Analysis for Waste Control

A. Accurate Waste Assessment

In chemical processing, aqueous waste streams contain varying amounts of dissolved solids as inorganic ions. To control total dissolved solids (TDS) loading into industrial waste removal facilities and to regulate treated waste effluent, periodic analysis is necessary. The standard method is to dry a filtered sample and weigh the solids residue after correcting for water of hydration, a more than two hour procedure and one that is quite difficult to automate. More frequent, on-stream determinations of TDS would enhance the reliability and efficiency of the waste treatment operation, provided that the data correlated sufficiently well with that of the standard method.

Direct conductivity monitoring appears to be an attractive alternative, but there is one difficulty. To determine TDS in grams per liter by means of conductivity, two calibration factors must be known and assumed constant: the weighted-average equivalent mass and the weighted-average ionic conductance. These combined assumptions easily lead to greater than 10% errors in TDS measurement using conductivity.

Process ion chromatography for complete sample analysis is certainly a possibility, but there might be a simpler approach. Total ionic content analysis (Small and Stevens, 1975a) is another flow injection–ion exchange method embodied into process instrumentation in order to monitor TDS in waste waters more accurately than can be done using direct conductivity. The complexity and maintenance needs of the total ionic content technique are in between those of in situ conductivity and ion chromatography.

B. *Exchange to Common Ions*

The total ionic content analysis scheme is depicted in Fig. 17. A simple water eluant conveys the injected sample to a cation exchange resin bed in the sodium form, where, again, in accordance with the principles of ion exchange (Inczedy, 1966; Wheaton and Seamster, 1966), sample cations remain on the exchanger while liberating an equivalent amount of sodium ions. Similar anion exchange to hydroxide occurs in the following column. Overall,

$$C_1^+ \ A_1^-, C_2^+ \ A_2^-, \ldots, C_n^+ \ A_n^- \quad \overset{R\text{–}Na^+, \ R\text{–}OH^-}{\Rightarrow} \quad Na^+OH^- \cdot \qquad (10)$$

where C_i^+ is the sample cation and A_i^- the sample anion.

Various sample ion species are exchanged for an equivalent proportion of single ion pairs. Detection and quantitation of the sodium hydroxide conductivity peak follows. By converting diverse sample ion species into only one ion pair type, conductivity measurement is rendered more truly representative of TDS, since knowledge of the changing weighted-average ionic conductance of the sample mixture is no longer needed. The same ion pair of known identity and conductance is detected in each analysis.

Direct comparisons show clear improvement in TDS measurement accuracy over direct conductivity using total ionic content. The difference in accuracy is most extreme for samples that may occasionally contain acids which are highly conductive but relatively low in TDS contribution. By converting hydrogen ions to sodium ions, the total ionic content analyzer reduces the conductance contribution of acids so that it is on a par with that of the sample salts.

Since total ionic content is not chromatographic in nature, column efficiency is less critical, and the ion exchange columns can be large enough to accommodate more than a month of repetitive sample injections.

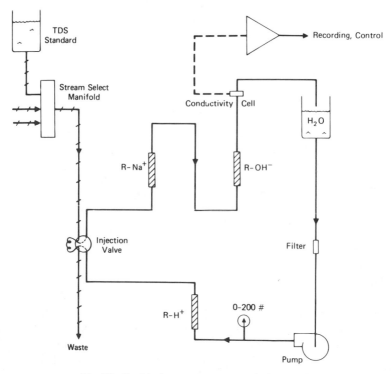

Fig. 17   Total ionic content process analysis system.

Periodic automatic admission of a TDS standard sample assures plant operators that the instrument is operating satisfactorily.

VI.   Differential Conductivity–Ion Exchange Method*

A.   *Alternatives for On-Line Acid or Base Analysis*

There are a variety of industrial processes wherein concentrated acid or base solutions play an important role. Examples are alkaline scrubber liquors for neutralization and removal of corrosive acid vent gases, electrolytic cell effluents, paper pulp cooking liquors, wastewater neutralizing additives, actual process feed reactants, and metal finishing solutions. In these instances acid or base concentration extends well above the levels appropriate for conventional pH meters since acidity ranges below pH 0 or alkalinity extends above pH 14. This

*See also Chapter 7.

generally corresponds to acid or base concentrations greater than 4% by weight and extending upwards to saturation.

On-stream titrators are commonly used for analysis of these higher concentrations of acid or base but generally require somewhat higher levels of maintenance. This is a consequence of the use of electrodes in titration combined with highly manipulative electromechanical components prone to wear and failure in continuous on-stream duty.

Another alternative for measuring and maintaining acid or base at these levels is the in situ conductivity probe equipped with a large cell-constant to accommodate elevated concentrations of acid or base. There is a problem with direct process conductance measurement, however: aqueous solutions exhibit conductivities that intrinsically plateau at intermediate concentrations of acid or base and begin, in fact, to *decrease* as levels continue upwards to saturation. This effect tends to limit direct conductivity systems to monitoring acid or base well below saturation.

Both conductivity and bulk density measurements for acid or base assay typically suffer from salt interference since it certainly contributes to solution conductivity and density.

These combined difficulties have led to efforts to explore new methods for on-line acid or base measurement.

B. Exchange to Water

Differential conductivity–ion exchange (DC–IX) is a new method for automated on-stream determination of concentrated acid or base concentrations (Miller and Stevens, 1980a,b). The technique is again based on flow-injection and ion exchange, followed by conductimetric detection. Like the total ionic content technique described previously, DC–IX is not at all chromatographic in nature. It is accurate to ±2% (Stevens and Miller, 1980); its principle advantage is reliability in continuous-duty applications, owing to the absence of chemical reagents, potentiometric electrodes, optical windows, and stirring mechanisms.

The analytical procedure is outlined in Fig. 18. A water eluant is again continually pumped through an injection valve that periodically admits a 5-μliter sample to the flowing stream. The injected acid or base next encounters a column packed with inert beads so as to spread or dilute the sample in the water carrier. Dilution here, typically fifty-fold, serves to spread reproducibly the sample zone to a range where conductivity is linear with concentration, as depicted earlier in Fig. 15. The dilution mechanism is an exaggerated "multiple path" or "eddy diffusion" effect regarded in chromatography as an undesirable peak-broadening process (Giddings, 1975). The effect is used to advantage here, however.

Next in the process, a conductance peak at cell 1 is monitored and stored.

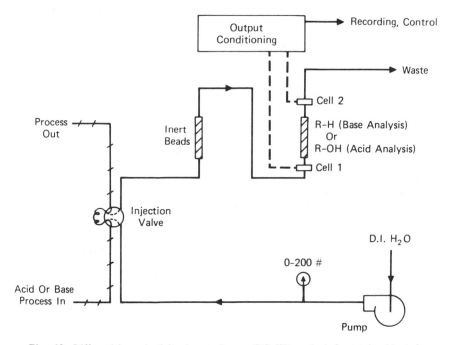

Fig. 18 Differential conductivity–ion exchange (DC–IX) method for total acid or base concentration.

In the following sequential step, the sample passes through an ion exchange column in the hydrogen ion form for total base assay or in hydroxide ion form for total acid determinations. For the sample acid or base constituent the following exchange occurs:

$$\text{Acids, } H^+\text{-}A^- \overset{\text{R–OH}^-}{\Rightarrow} H_2O \tag{11}$$

$$\text{Bases, } C^+\text{-}OH^- \overset{\text{R–H}^+}{\Rightarrow} H_2O \tag{12}$$

where R again denotes stationary ion exchange particles retained in the column.

The key to this process is that the acid or base of analytical interest is converted to water and thus rendered indistinguishable conductimetrically from the water eluant.

A conductance peak is again read and stored from cell 2, since the difference in cell conductivity measurements is proportional to the acid or base content of the original sample.

A complication arises in that salt from the process sample is also converted to another form by ion exchange,

$$\text{Acids, } C^+A^- \overset{\text{R–OH}^-}{\Rightarrow} C^+OH^- \tag{13}$$

$$\text{Bases, } C^+A^- \overset{R-H^+}{\Rightarrow} H^+A^- \tag{14}$$

requiring a reduction in the sensitivity of the second cell to achieve equal responses from both cells for a representative background salt mixture.

Until replaced or regenerated automatically the ion exchange column will accommodate N sample injections where

$$N = \frac{C_r \times V_c}{C_s \times V_s} \tag{15}$$

and C_r is the resin exchange capacity, V_c the exchange column volume, C_s the average sample concentration, and V_s the sample injection volume.

This equation accurately predicts more than two weeks of operation with continuous 15-min injections before replacement or regeneration is warranted.

C. The Process Instrument

The DC–IX analyzer is pictured in Fig. 19. Fluid-handling components are all confined to the lower section to protect electrical equipment and facilitate clean-up. The pneumatically actuated sample injection valve appears to the lower right of the bottom compartment. To the left of the injection valve are the dilution and ion exchange columns, and to the extreme left are mounted the miniature flow-through conductivity cells. This lower panel is fabricated of plastic to eliminate the corrosive effects of inevitable sample acid or base spills. The graphic flow pattern is etched into a laminated plastic overlay and significantly aids in assembly, check-out, and trouble-shooting in field operation.

The panels viewed in Fig. 19 are also hinged to afford convenient front access to regulators, valves, and the pump mounted inside.

Electronics are housed in the separate air-purged, gasketed upper portion of the cabinet.

A DC–IX analyzer engaged in process analysis appears in Fig. 20. In the application shown, the plant has conserved the amount of neutralizing base used in an acid gas scrubber system. Note the bypass filter arrangement for sample entering the lower right hand corner of the instrument. Approximately centered in the figure is a double-block-and-bleed valve system for admitting purge water, facilitating safe maintenance as described in Section II above.

VII. Summary

Successful on-stream process analysis is based upon adequate sample preparation.

Automated flow injection analysis methods exploiting newly improved liquid

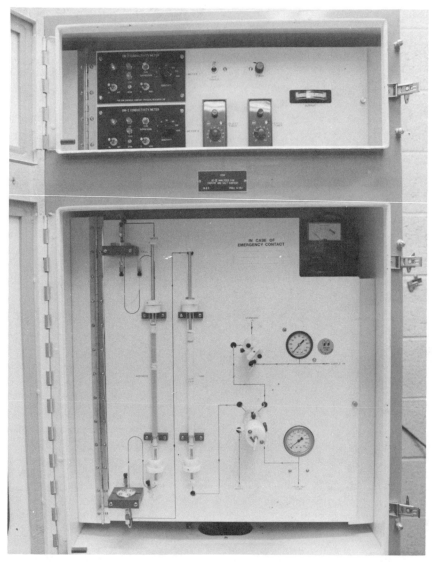

Fig. 19 DC–IX process analyzer. [Reprinted with the permission of the Instrument Society of America, 1980, from articles appearing in *Adv. in Instr.* **33–35.**]

Fig. 20 Process DC–IX analyzer in alkaline vent scrubber application.

chromatography components are playing an expanding role in chemical process control. Ion exchange columns in different arrangements have enabled versatile on-stream flow-injection process analysis of a wide variety of samples:

(a)   ion chromatographic determination of dozens of ionic species to parts-per-million levels,

(b)   ion exclusion chromatographic analysis of trace acids in concentrated process brines,

(c)   total ionic content assay of waste waters, and

(d)   total acid or base measurement.

The ion exchange techniques above, along with the dilution and extraction sample-conditioning methods outlined earlier, are patented by The Dow Chemical Company. (These patents are listed within the references.) Information regarding licensing of this technology may be obtained from the Patent Department, 1776 Building, The Dow Chemical Company, Midland, Michigan 48640. The Dionex Corporation of Sunnyvale, California, offers ion chromatography products for sale under license from The Dow Chemical Company.

ACKNOWLEDGMENTS

A number of individuals have contributed substantially to the subjects described in this chapter. Hamish Small and Timothy Stevens, co-inventors of ion chromatography, and Virgil Turkelson, Thomas Peters, and H.D. Ruhl of the Analytical Laboratories, Michigan Division, The Dow Chemical Company, all played significant roles in defining and refining the analytical procedures that appear here. In addition, numerous instrument engineering and design inputs originated from William Parth, Paul Hargash, Kenneth Cabala, and William Albe as employees of the Instrument Applications Laboratory, Michigan Division, Dow Chemical Company.

References

Bell, D. R., and Miller, T. E., Jr. (1980). U.S. Patent pending.

Borman, S. A. (1980). *Anal. Chem.* **52**(13), 1409A–1410A.

Daniels, F., and Alberty, R. A. (1966). "Physical Chemistry," 3rd ed., p. 5. Wiley, New York.

Giddings, J. C. (1975). *In* "Chromatography" (E. Heftman, ed.), 3rd ed., p. 32. Van Nostrand-Reinhold, Princeton, New Jersey.

Inczedy, J. (1966). "Analytical Applications of Ion Exchangers," 1st ed., p. 161. Pergamon, Oxford.

Maugh, T. H. (1980). *Science* **208**, 164–208.

Miller, T. E., Jr. (1978). *ISA Trans.* **18**(2), 59–64.

Miller, T. E., Jr., and Cabala, K. M. (1979). U.S. Patent 4,148,610.

Miller, T. E., Jr., and Stevens, T. S. (1980a). *Adv. Instrum.* **35**, 21–27.

Miller, T. E., Jr., and Stevens, T. S. (1980b). U.S. Patent 4,199,323.

Mulik, J. D., and Sawicki, E. (1979). "Ion Chromatographic Analysis of Environmental Pollutants." Ann Arbor Sci. Publ., Ann Arbor, Michigan.

Richards, M. (1975). *J. Chromatog.* **115**, 259–261.

Small, H., and Solc, J. (1976). *In* "The Theory and Practice of Ion Exchange" (M. Streat, ed.), pp. 32-1–32-10. Soc. Chem. Ind., London.

Small, H., and Stevens, T. S. (1975a). U.S. Patent 3,918,906.

Small, H., and Stevens, T. S. (1975b). U.S. Patent 3,920,398.

Small, H., and Stevens, T. S. (1978). U.S. Patent 4,101,460.

Small, H. Stevens, T. S., and Bauman, W. C. (1975). *Anal. Chem.* **47**(11), 1801–1809.

Snyder, L. R., and Kirkland, J. J. (1979). "Introduction to Modern Liquid Chromatography," 2nd ed., p. 111. Wiley, New York.

Stevens, T. S., and Miller, T. E., Jr. (1980). *Anal. Chem.* **52**(13), 2023–2026.

Stevens, T. S., and Small, H. (1978). U.S. Patent 3,966,596.

Stevens, T. S., Turkelson, V. T., and Albe, W. R. (1977). *Anal. Chem.* **49**(8), 1176–1178.

Wheaton, R. M., and Seamster, A. H. (1966). *Kirk-Othmer Encycl. Chem. Technol., 2nd ed., 1963–1971* **11**, 875.

2

Flow-Injection Analysis: A New Approach to Near-Real-Time Process Monitoring

CRAIG B. RANGER

QuikChem™ Systems Division
Lachat Chemicals, Inc.
Mequon, Wisconsin

I. Introduction

Continuous monitoring has been demonstrated to be a highly cost-effective technique for determining multiple analytes in flowing streams of diverse sample matrices. Its utility and economy have improved consistently with the development of new analytical methods and electronics. Evolution has occurred from the relatively simple pH electrode through wet chemical analyzers and chromatographs to provide systems that can determine several analytes simultaneously in a complex sample matrix. These monitoring capabilities have formed the basis by which further information processing can effect control functions and facilitate automated process optimization.

The purpose of this chapter is to discuss a relatively new analytical technique known as flow-injection analysis (FIA) applied to automated process stream analysis. FIA will be defined and discussed in terms of the operating principles, the various possible configurations, and its applications to on-line monitoring. The advantage of FIA over segmented flow analysis (SFA) for process monitoring will be stressed.

Flow-injection analysis is based on the repetitive injection of aliquots into a continuously flowing, unsegmented stream. Following injection the aliquot may be delivered directly to a detector or dispersed in a carrier stream to facilitate sample conditioning prior to measurement. The ability to accurately and precisely establish a wide range of dispersion conditions in the system is the key to the operation and versatility of FIA.

Modern flow-injection analysis was introduced in Hungary during 1970 by Nagy *et al.* Their initial purpose was to use sample injection into a carrier stream as a means to deliver reproducible sample volumes to an electrode following stirred mixing in order to effect high analytical rates. They found the analytical readout to be in the form of transient peaks with baseline resolution between samples as illustrated in Fig. 1.

Expanding on his work, Stewart *et al.* (1976) in the United States and Ruzicka and Hansen (1975) in Denmark simultaneously modified the technique. Their major discovery was that dispersion of the injected sample into the carrier stream could be induced by the flow process alone and did not require any mechanical assistance. Furthermore, it was found that the dispersion in narrow bore tubes could be controlled so as to effect various degrees of mixing between sample and carrier solution and could be maintained beneath an acceptable upper limit to prevent excessive dilution. Figure 2 shows a generalized schematic for the FIA process.

These were significant findings, since it had long been thought that air bubbles had to be introduced at frequent and regular intervals in order for a continuous flow analyzer to work. The bubble was purportedly necessary to limit sample dispersion and generate turbulent flow to promote mixing and to scrub the walls of the analytical conduits in order to minimize carry-over and cross-contamination between samples. In fact, all three of these functions can be implemented in an unsegmented stream.

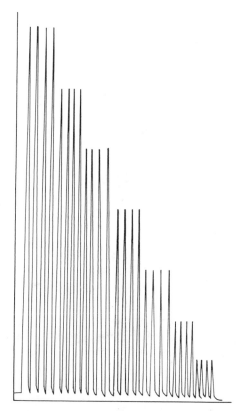

Fig. 1. Typical recorder output of transient peaks associated with FIA limited and medium dispersion conditions.

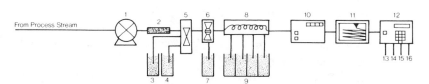

Fig. 2. Generalized schematic of Lachat QuikChem™ process monitor: (1) sample suction pump, (2) in-line filter, (3) standard solution, (4) rinsing solution, (5) 3-way solenoid valve, (6) sample injection valve, (7) primary reagent or diluent, (8) reaction manifold, (9) additional reagents, (10) detector, (11) recorder, (12) microprocessor, (13) analog or digital data transmission line, (14) hi–lo alarm, (15) automatic drift correction, and (16) automatic recalibration.

TABLE I

COMPARISON OF CHARACTERISTICS OF FIA AND SFA

Parameter	FIA	SFA
Sample introduction	Injection	Aspiration
Sample volume	Small (μliter)	Large (mliter)
Analytical stream	Unsegmented	Segmented
Manifold conduits	< 1-mm i.d.	2-mm i.d.
Lag phase	Negligible	Significant
Pump speed	Variable	Fixed
Pumping pressures	Low	Low
Column	Possible	Possible
Data reduction	Integration or peak height	Peak height

II. Principles

A. Flow-Injection Analysis versus Segmented Flow Analysis

Although certainly interesting, such a finding would not be too important unless analysis in an unsegmented stream had significant practical advantages over the well-established segmented flow paradigm. In fact, the several limitations which the air bubble places on the capabilities of a continuous flow analyzer are well known, and they can all be circumvented by flow-injection analysis. A comparison of the operating characteristics of the two techniques is presented in Table I.

In order for a continuous flow analyzer to work there are two critical factors that must be considered. First, a mechanism must be present to effectively and reproducibly mix the sample with the reagents. Second, the volumetric integrity of the sample must be maintained to prevent excessive dilution and consequent quelling of the response signal.

Segmented flow analysis addresses these two requirements by dividing the analytical stream into a large number of boluses with air bubbles. Each section bounded by two air bubbles forms, in effect, a reaction chamber. Some intermixing of the segments does occur through limited dispersion and carry-over owing to deposits on the inner walls of the reaction coils. However, excessive sample dispersion and, therefore, dilution is prevented. In addition, a mixing action is performed within each segment as turbulent flow is generated by the interaction of the bubbles with the carrier stream and coil walls. A wash solution is introduced at defined intervals to reestablish baseline and check for drift.

B. Dispersion Processes

In flow-injection analysis, the injected sample disperses into and thereby mixes with the analytical stream under laminar flow conditions. The use of narrow bore tubing as the reaction conduit is the key to preventing massive sample dispersion. In fact, by varying the operating parameter values of an FIA system, including sample volume, flow rate, tube length, and tube diameter, a wide range of dispersion conditions can be established to attain optimum conditions for a particular analytical technique. It is this controlled dispersion that allows FIA to operate in diverse situations.

The fundamental questions remaining are how mixing is actually carried out and how excessive deposition on the conduit walls is prevented under the laminar flow conditions used in FIA. According to chemical reactor engineering theory as presented by Taylor (1953), laminar flow in open bore tubular reactors results in plug flow whereby, among other things, the linear velocity of the flowing stream is zero at the liquid–wall interface. If this were universally true, the excellent short-term baseline resolution exhibited in FIA would be impossible.

Although the exact mechanism of action of mass transport under FIA conditions is not entirely understood, Vanderslice *et al.* (1981) have proposed a first level quantitative model which is sufficiently well developed to justify review. They have contributed significantly to our understanding of the dispersion process and development of basic guidelines to designing FIA manifolds.

Their most important fundamental discovery was that the published work on FIA had not been carried out in the laminar flow regions described by Taylor. His equations solved for situations in chemical reactor engineering where either diffusion or convection is dominant in the mixing process. In fact, a combination of diffusion and convection processes appears to operate. This explains the previous discrepancies between early FIA theory as proposed by Ruzicka and Hansen (1978) and actual experimental data.

Specifically, two important principles were elucidated. First, it was found that axial sample dispersion does not increase ad infinitum under FIA conditions, as it should if FIA operated in the Taylor regions. This finding has at least two important practical implications: (1) FIA operating conditions can be established whereby a high flow rate is used to process samples without encountering undesirable dispersion and (2) limited dispersion conditions can be attained even in long analytical lines without causing excessive sample dilution. This latter attribute was thought to be strictly the province of segmented flow analysis.

The second principle provides an explanation for the minimal carry-over and cross-contamination experienced with FIA. According to their calculations, axial molecular diffusion is relatively inactive compared to radial molecular diffusion.

This mechanism acts to move molecules back and forth between the tube center and wall, thus serving as a scrubbing mechanism.

Vanderslice's group also developed two equations that describe the sample travel time from injection valve to detector and baseline-to-baseline time for the peak. As was mentioned above, there are four variables that affect the dispersion conditions in an FIA system—sample volume, flow rate, tube length, and tube diameter. The Vanderslice equations include these as well as two constants: one for the molecular diffusion coefficient of the sample and one to account for variability in sample concentration and detector sensitivity. They are as follows:

$$t_a = \frac{109a^2 D^{0.025}}{f} \frac{L}{q} , \tag{1}$$

$$t_b = \frac{(35.4)a^2 f}{D^{0.36}} \left(\frac{L}{q} \right)^{0.64}, \tag{2}$$

where t_a is the travel time, t_b, the baseline-to-baseline time, a, the tube radius, D, the molecular diffusion coefficient, L, the tube length, q, the flow rate, and f, the concentration/sensitivity constant.

These equations are helpful in the initial design of an FIA manifold to establish proper operating conditions for a given dwell time and analytical rate.

Flow-injection analysis is most popularly described in terms of the three dispersion categories of limited, medium, and large, each of which is used for different purposes. These three dispersion types are illustrated in Fig. 3. Limited dispersion is attained by injecting a relatively large sample volume into a short length of tubing which connects directly to a detector. In this way band spreading is minimized, and the readout is in the form of highly transient peaks. A large number of samples can thus be processed per unit time using any detector having flow-through capability.

Medium dispersion is used to generate partial mixing of the sample and reagents. The majority of flow-injection analyses are carried out in this dispersion range to perform a host of colorimetric, fluorimetric and electrochemical reactions. Several associated sample treatment functions can also be performed in-line, including dilution, solvent extraction, dialysis, ion exchange, and oxygen scrubbing.

Large dispersion has been used primarily to perform continuous flow titrations. The technique was originally described by Ruzicka *et al.* (1977) and subsequently automated by Stewart and Rosenfeld (1981). In this case a fixed volume of sample is injected and thoroughly mixed with titrant in a mechanically stirred chamber to form an exponential concentration gradient. A detailed explanation of FIA titrimetry will be presented in the section on special techniques.

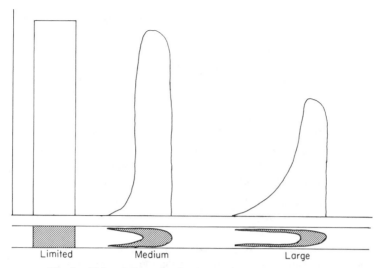

Fig. 3. FIA peak forms for the three major dispersion categories.

As was mentioned previously, FIA utilizes transient signals which represent some percentage of the steady-state condition established in SFA. Therefore, since the result is taken from the exponentially rising portion of the peak, it is vital that reproducible sample volumes be introduced into the system and that each sample remain in the analytical stream for the same amount of time. Any variability in these parameters will cause differences in dispersion and reaction times which will be reflected in the overall precision of the system.

III. Features

A. *Stability*

One of the major problems of segmented flow analysis is establishing and maintaining a stable flow pattern. The situation is improved by regularly metering in a precise volume of air and adding a surfactant to the liquid stream to reduce interfacial tension between the stream and the inner wall of the glass holding coil. However, even with these measures, smooth and reproducible flow are often not attained for about 30 min after system startup.

In contrast, a single phase stream stabilizes immediately after flow is initiated. No mechanism is required for air injection, debubbling, or resampling, and surfactants need not be introduced. Furthermore, only an insignificant amount of pumping pressure is lost over the length of the analytical line in FIA. This reduced head pressure ensures stable flow rates and obviates the requirement of repumping the solution from the flow cell, which is standard practice in SFA. Thus, FIA facilitates prompt system startup and intermittent pumping, which can be used to conserve reagents, increase reaction time, and perform kinetic assays.

B. Stopped-Flow Capability

Fast system startup is desirable for a process monitor. Primarily it allows the unit to be turned on and off as required. This means that the system need only be running when the process stream is actually being analyzed. The result is decreased reagent consumption and reduced maintenance on the entire analyzer. And, less significantly, when maintenance is required the system can be restarted quickly.

In addition, the flow can be stopped during the actual analysis, which may be desirable if a reaction is rate-limited. Incubation time can be extended without consuming unnecessary reagent volumes or using long holding coils. Stopped flow can also be used to carry out kinetic analyses. The excellent stability of an unsegmented stream allows it to be stopped reproducibly. As a result, the reaction product can be pumped into the flow cell of a detector and stopped there to determine a reaction rate. Alternatively, a spectrophotometric or voltammetric scan can be performed to, for example, determine several analytes simultaneously.

C. Response Time

Of particular importance in process monitoring is the response time of the analyzer. That is the time from sample input to readout. Of the more than 50 industrial methods developed to date in FIA, none has a response time of greater than one minute and most are around 30 sec. However, the limiting time in FIA is not the response time, but rather the time required for a single sample to be thoroughly processed by the monitor. If multiple samples are present in the analytical line, the sample stream can be analyzed at a rate of 3 to 6 times per minute, thus, near-real-time wet chemical monitoring becomes possible. This contrasts with segmented flow response times, which fall in the range of 5 to 20 min.

The short response time also allows for frequent monitoring of the actual analytical conditions. The process is generally initiated by introducing a blank sample into the system and correcting for any baseline drift. A standard is then introduced to recalibrate the system after which sampling is resumed. This entire process takes a maximum of 2 min versus a minimum of 10 min for SFA. These

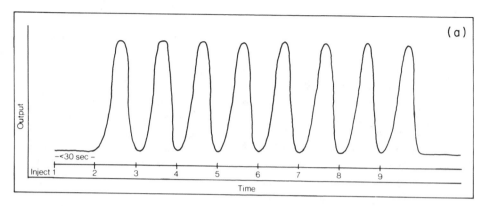

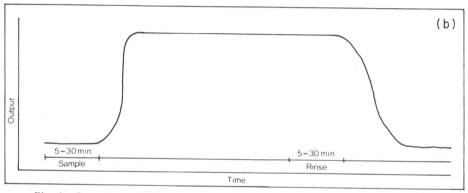

Fig. 4. Comparison of FIA (a) and SFA (b) output forms with associated nominal response times.

response times are indicated in Fig. 4, which also illustrates the differences in monitor output between SFA and FIA. Although it is clear that a good process monitor will be a stable instrument remaining substantially free from baseline and sensitivity drift, it nevertheless is valuable to be able to confirm its accuracy and precision automatically at frequent intervals.

IV. Techniques

A. *Dilution*

One of the important aspects of a continuous flow analyzer is its capability to pretreat samples in-line. Flow-injection analysis not only performs several of these functions, but also does them faster and more reproducibly than other methods.

If the actual concentration of an analyte causes the effective range of the chemistry or detector to be exceeded or if substantial interferences are present, the sample can be *diluted*. This is generally effected in segmented flow analysis by combining a segmented sample stream with the diluent and pumping it through a holding coil. The air bubbles are then removed and immediately reinjected to ensure a regular flow pattern. Dialysis can also be used for dilution although this method is usually reserved for removing interfering macromolecules.

FIA facilitates three dilution methods. The first was originally identified by Ruzicka and Hansen (1978) whereby the injected sample volume is simply reduced. As they stated, since the rising portion of the analytical peak is approximately linear, decreasing the sample volume will reduce the peak height nearly quantitatively. This method is the most straightforward but is constrained by a lower limit defined by the minimum sample volume accommodated by the injection valve.

This limitation was of less concern in the early stages of FIA development, when relatively large sample volumes were being used. Thus ten and one-hundred-fold dilutions could be made by reducing a standard 500-μliter injection volume to 50 and 5 μliter, respectively. However, now that smaller standard sample volumes, on the order of 50 μliter, are being used to minimize flow rate and reagent consumption, this method becomes impractical.

One obvious alternative is to inject a minimum sample volume and carry it through a mixing coil of increased diameter and/or length. Using this procedure concomitantly with dilution will result in higher hydrodynamic pressure, increased sample dispersion (band spreading) and a longer residence time; the last two contribute to an undesirable reduction in the overall analytical rate.

Reis *et al.* (1981) have proposed another technique, elegant in its simplicity and effects, which they call the zone-sampling process. In this case, two independent flow circuits are used, as illustrated in Fig. 5. The sample is injected into the first stream and pumped through a coil to effect some degree of dispersion. This stream itself is pumped through a second injection valve, which is switched at an appropriate time to capture and inject a portion of the dispersed sample zone into a second carrier stream. Although this may at first appear reminiscent of the dilution loop used in SFA, it is far from it. The initial dispersion is effected quickly here and without the difficulty of injecting, removing, and reinjecting air bubbles. Furthermore, although two injection valves are required, the mechanism is simple and economical, for example, the injector–commutator of Bergamin *et al.* (1978) or the dual rotary valve assembly of Ranger (Lachat Data Sheet 1000-600) and shown in Figs. 6a and 6b.

Reis's group has also pointed out three additional advantages of zone sampling. First, this process can be used to effectively split the sample stream for multichannel (simultaneous) analyses where different dilution ratios or dispersion conditions are required. Second, one standard can be injected repeatedly

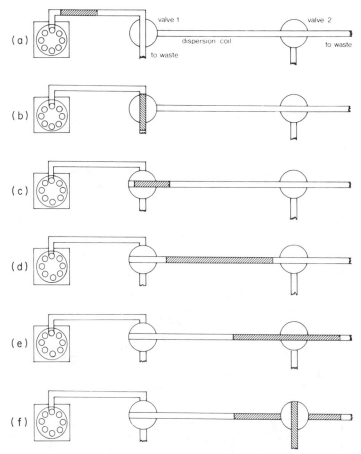

Fig. 5. Zone-sampling process: (a) aspirate sample, (b) load injection valve 1, (c) inject sample, (d) disperse sample, (e) load valve 2, (f) inject portion of dispersed zone.

through the first valve, and then selected portions of that dispersed sample can be sequentially injected through the second valve to generate a calibration series. This is accomplished by simply switching the second valve at different intervals. Finally, they have suggested that the dispersion of the sample zone can be thoroughly mapped by sequentially injecting each portion and observing the pattern generated on the recorder.

B. Solvent Extraction

There are many cases that exhibit the opposite problem of that previously described where the sample must be concentrated or removed from a matrix

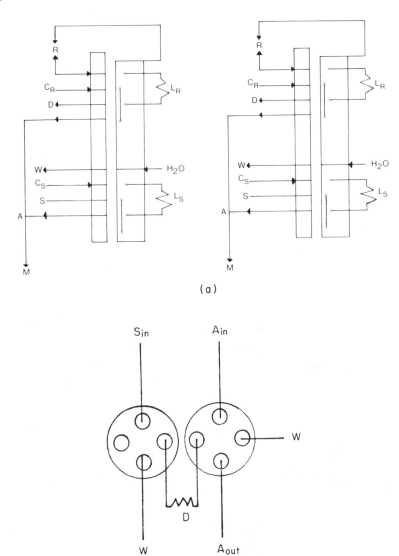

Fig. 6. (a) Schematic presentation of the double proportional injector in the loading (A) and injection (B) positions. S and R are sample and reagent; L_S and L_R are the sample and reagent loops; C_S and C_R are the sample and reagent carrier streams; D is the reagent out-flow to a recovery vessel; W is the waste; A is the confluence point placed 5 cm from the injection ports; M represents the manifold. [Reprinted with permission from Bergamin *et al.* (1978).] (b) Dual rotary valve arrangement for performing merging zones and zone dispersion techniques. S_{in}, sample in; W, waste; D, dilution coil; A_{in}, analytical stream in; A_{out}, analytical stream out.

containing interferences. FIA can accomodate several methods for accomplishing this. One of the most powerful and versatile of these is solvent extraction.

Solvent extraction can perform three functions: (1) separation of the analyte from a turbid sample, an interfering dye or a biological matrix such as blood or urine; (2) dissolution of a colorimetric or fluorimetric species in the organic phase, which is otherwise insoluble in the aqueous phase and would therefore give a poor readout as it passes through the detector; (3) concentration—a significant percentage of analyte can be removed into a relatively small volume of solvent.

It is well established that a solute will become distributed between two immiscible liquids such that the ratio of thermodynamic activity of the species in each phase is constant. This distribution law may be written as

$$A_1/A_2 = K_d \tag{3}$$

where A_1 is the solute activity in Phase 1, A_2 the solute activity in Phase 2, and K the distribution coefficient of solute A.

Several extraction procedures have been automated using flow injection. It is interesting to note that in this application the flow-injection system effectively becomes a segmented flow analyzer. The aqueous and organic phases are combined into regular alternating segments through a phase combination (PC) fitting first described by Karlberg and Thelander (1978) for FIA. However, it should be noted that Wallace demonstrated in 1967 the feasibility and desirability of this approach using an AutoAnalyzer without air segmentation.

The Karlberg PC fitting consists of a tee into which two concentrically arranged lengths of PTFE tubing are inserted. The two phases are introduced through the two inlet arms of the tee. A regular two-phase pattern is established as they flow through the PTFE tube, and, in fact, the lengths of the segments can be changed by adjusting the axial position of the inner tub. Kawase has also reported a segmentor whose phase ratio is changed by locating the inlets at different relative angles (Kawase *et al.*, 1979).

Once the segmented pattern is attained the stream is pumped through an extraction coil. Bolus flow is promoted wherein the interface between the segments is continuously refreshed, thereby maintaining a maximum diffusion rate resulting in an efficient extraction process. By dividing the aliquot into small portions a large surface area is generated and the emulsions often produced in manual extractions are avoided. Using this method ion-pair extractions, metal chelate extractions, and determinations of extraction constants have all been performed using FIA.

It is vital to the ultimate success of an automated solvent extraction that there be dependable separation of the phases prior to delivery to a detector. Karlberg described a T-fitting for this purpose one leg of which was threaded with Teflon strands to bias the flow of organic phase (Karlberg and Thelander, 1978). However, the use of a degassed and chilled solvent was necessary; the aqueous phase had to be carefully controlled to prevent its entry into the flow cell, and the effectiveness of the separator was sensitive to solvent concentration (Kawase *et al.*, 1979). Kawase *et al.* (1979) and Karlberg (1980) therefore independently introduced phase separators incorporating a PTFE membrane, which is selectively permeable to the organic phase. These units illustrated in Fig. 7 appear to operate more dependably and facilitate high sample throughout.

The improvement of analytical rate over that possible with an air-segmented system is significant. For example, Kawase first reported an air-segmented system for ion-pair extraction of cationic surfactants operating at 10–20 samples per hour (Kawase and Yamanaka, 1979). He subsequently described a flow-injection approach that processed 60 samples per hour (Kawase *et al.*, 1979).

The simplicity of the FIA method relative to the SFA approach is also easily seen by referring to the respective manifold schematics given in Fig. 8.

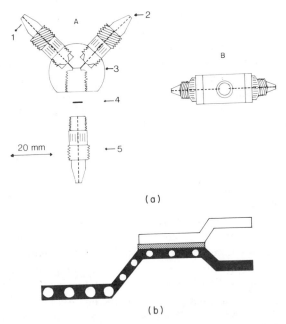

Fig. 7. (A) Construction of a phase separator: (A) Side view; (B) bottom view of PTFE body; (1)/(2) PTFE joints (inlet/outlet); (3) PTFE body (axial pitched cylindrical cavity); (4) PTFE porous membrane sheet (7-mm diameter); (5) PTFE joint (outlet to flow cell). Inside diameter of the capillaries of both joints and body is 0.8 mm. The inner volume of the separator is 12.5 μliter. $\bigcirc$, PTFE LC three-way joint. $\square$, aqueous phase; $\boxtimes$, organic phase; $\blacksquare$, PTFE membrane. (B) PTFE membrane is sandwiched between two acrylic blocks with flow channels cut into the faces.

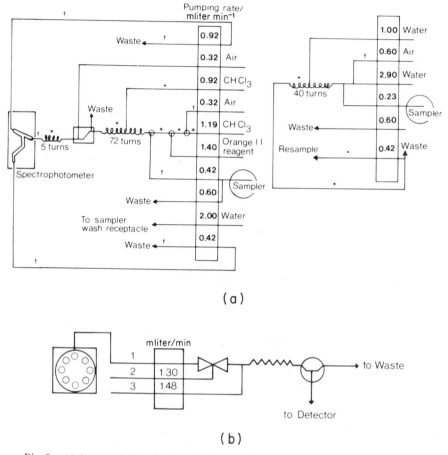

Fig. 8. (a) Automated determination of cationic surfactants using segmented flow analysis. (b) Automated determination of cationic surfactants using flow-injection analysis.

Finally, an example of the excellent quantitative aspect of solvent extraction by FIA is presented in Fig. 9. Both chromatograms were run on polynuclear aromatic hydrocarbons extracted from oil (Shelly, 1982). As is evident, the FIA extraction demonstrates an excellent correlation with the manual extraction.

C. Dialysis

Another important technique to separate an analyte from interferences is *thin-film dialysis*. Dialysis involves separation of solution components based upon differential diffusion rates through a semipermeable membrane. In continuous

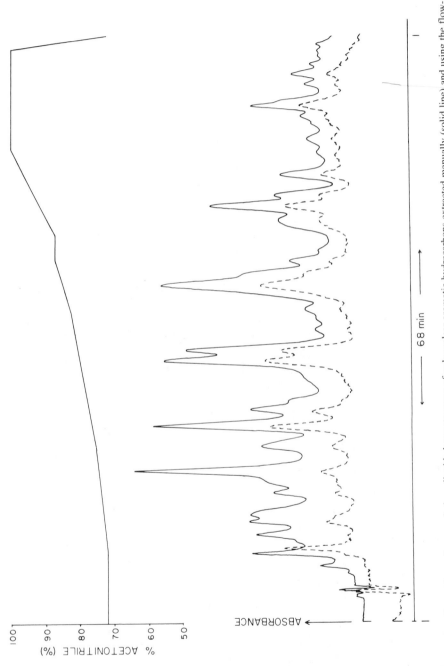

Fig. 9. Comparison of reversed phase liquid chromatograms of polynuclear aromatic hydrocarbons extracted manually (solid line) and using the flow-injection technique (dotted line). The associated gradient profile is shown at top.

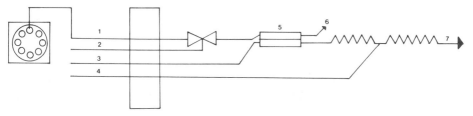

Fig. 10. Automated determination of calcium in milk samples using flow-injection analysis with dialysis: (1) sample, (2) H$_2$O, (3) CPC, (4) AMP, (5) dialyzer, (6) to waste, (7) to detector.

flow systems this is accomplished by positioning a membrane between two acrylic blocks, each having a semicircular channel cut into it.

The type of membrane used is dependent upon the species to be dialyzed. Typically, a cellulose construction will be used to dialyze macromolecules such as proteins and fats. This technique has been applied to blood, urine, milk, meat, and various other food samples.

For example, Basson (1979) has reported the determination of calcium in milk samples following dialysis. The system shown in Fig. 10 analyzes 180 samples per hour with repeatability of better than 1%.

Another method is gas dialysis. In this case a gas is generated in the donor stream, which diffuses across a membrane of silicone or Teflon into a recipient stream. The volatile analyte may thus be removed from a nonvolatile sample matrix. Karlberg and Anfalt (1980) have applied this technique to the determination of ammonia (Fig. 11) whereby they attained an analytical rate of up to 90 samples per hour and a detection limit of 0.02 ppm.

D. Column Treatment

Columns have been incorporated into flow-injection systems for several purposes. That this is possible using an unsegmented stream without encountering severe band spreading is not surprising in view of the success of high-performance liquid chromatography.

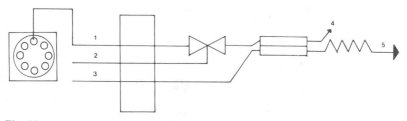

Fig. 11. Automated determination of ammonia in water samples using flow-injection analysis with gas dialysis: (1) sample, (2) NaOH, (3) phenol red, (4) to waste, (5) to detector.

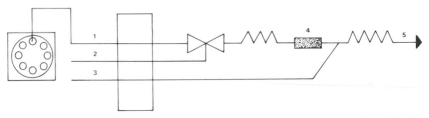

Fig. 12. Automated determination of nitrate and nitrite in water samples using flow-injection analysis with column treatment: (1) sample, (2) NH₄Cl, (3) color reagent, (4) column, (5) to detector.

The cadmium reduction method for converting nitrate to nitrite followed by the classical diazotization/coupling reaction with N-1-naphthylethylenediamine dihydrochloride and sulfanilamide has been automated using FIA. Here the sample is injected into ammonium chloride buffer and pumped through a short column packed with copperized cadmium granules. The reduction product is then combined with the color reagent carried through a mixing coil and delivered to a spectrophotometer. This method, which analyzes 120 samples per hour with better than 1% reproducibility, is shown schematically in Fig. 12. It is important to note that even with a column in-line, the lag time from injection to readout is only 20 sec.

The sample loop of an injection valve can be replaced by columns packed with various materials to separate the analyte from interferences prior to injection. There are two cleanup methods that can be employed, as illustrated in Fig. 13. In the first case, where the column has a greater affinity for the interferences the column is loaded and the valve is switched to the inject position. The sample is injected for a predetermined time period, allowing the carrier stream to elute the analyte. The valve is then switched back, and an elution solvent is directed through the column to wash out the interferences and carry them off to waste. A conditioning solvent may then be directed to the column to prepare it for the next sample.

Alternatively, if the analyte is retained longer on the column than the interferences, the sample is aspirated through the column for a length of time sufficient to pass the interferences through. The valve is then switched, and the isolated analyte is injected. If other interferences are present with longer retention times, the valve can be switched back to the load position, as described in the previous paragraph.

If the analyte is strongly retained on the column, the second sample pretreatment method can be used as a concentration step. The sample is aspirated through the loop, and a zone of analyte accumulates on the column while the interferences pass on to waste.

All of the above steps can be automated with simple electronics or by using a

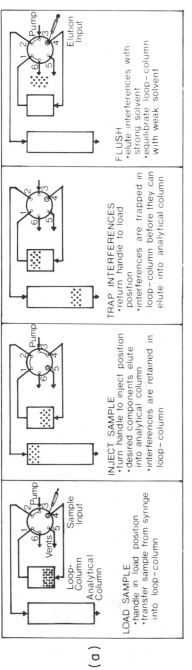

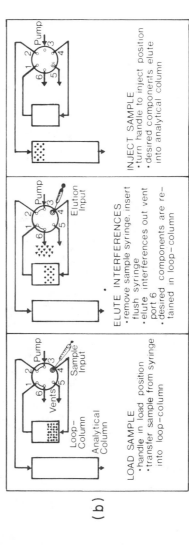

Fig. 13. Schematic presentation of loop-column function for sample cleanup when interfering compounds are (a) more and (b) less strongly retained than the analyte. [Courtesy of Rheodyne, Inc. (1981).]

microprocessor. Bergamin *et al.* (1980) were the first to apply this technique to flow-injection analysis. They replaced the valve sample loop with an Amberlite IR-120 ion exchange resin to remove several interfering cations from rain water samples to increase the sensitivity of ammonia determination using Nessler's reagent. The sample was aspirated through the column, the interferences passing through to waste. Upon injection the sample was eluted from the column with sodium hydroxide and combined with the color reagent. An analysis rate of 40 samples per hour was achieved with 2% precision, 200-μg/liter detection limit, and nearly 100% sample recovery.

It should be emphasized that the columns described above are short and loosely packed. As a result, low-pressure pumps can be used to move liquid through them. In addition, they can be packed with a wide variety of materials and even used in series to facilitate design of a highly specific sample pretreatment system.

E. *Electrochemical Scrubbing*

Having now covered the various sample treatment functions of general interest, I shall consider one which addresses a specific problem. In recent years electrochemical detectors have become popular for use with liquid chromatography (LCEC). The chromatography serves as a sophisticated sample preparation system, and electrochemical detection provides high sensitivity and specificity, thus expanding the system capabilities beyond those possible with UV, RI, and fluorescence detectors. However, there are many species that can be detected in a sample matrix that contains no significant interferences. In addition, an electrochemical system containing potential interferences can often be manipulated by selecting buffer potentials of interfering molecules away from that of the analyte. Solvent extraction and cleanup can also be applied. Thus, the time-consuming and relatively expensive chromatography can become unnecessary and can be replaced by flow-injection analysis.

The LCEC literature describes many systems that use oxidative detection. However, reductive detection has been relatively unpopular, primarily due to the omnipresence of oxygen, which significantly decreases sensitivity and increases ambient noise, since it causes high background currents. Reducible organics and metal ions also interfere. Therefore, some interest in solving this problem has been generated. Several approaches have been used in removing oxygen interference, such as nitrogen sparging, reverse pulse amperometry, and chemical reduction. None of these techniques, however, is easily applied to process monitoring systems since they are relatively complex or inflexible.

Recently an electrochemical scrubber that substantially removes interferences

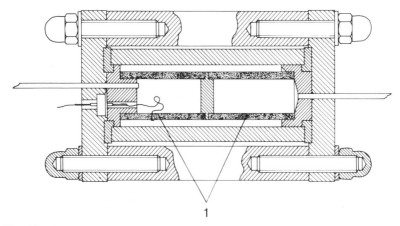

Fig. 14. Continuous electrochemical scrubber. As analytical stream passes through the unit, interfering species, such as oxygen, metals, and organics, are reduced at the porous silver electrodes—1.

due to oxygen, organics, and metals was developed (Hanekamp *et al.*, 1980). Originally intended for HPLC, the apparatus is also applicable to FIA.

The scrubber is based on the removal of reducible impurities, using two sequential porous silver electrodes. The device, illustrated in Fig. 14, was designed to accommodate flow rates sufficiently high to facilitate fast scrubbing so as not to be rate limiting.

In order to estimate the optimum scrubbing potential, a voltammogram is run on the carrier stream. Figure 15 shows a voltammogram run on a buffer stream before (a) and after (b) addition of methanol to the system. A plot of the background current versus scrubbing potential at two detector settings indicates a dramatic decrease of current when the scrubber is operated in the range of -1.4 to -1.5 V.

The original study on the efficacy of the scrubber was performed at a dropping mercury electrode. From Table II it may be seen that the background current reflected in the baseline values as well as noise are significantly decreased below the ambient value of 1 μA obtained by using nitrogen purging alone.

Once the sample has been pretreated using one or more of the methods previously described, it can be either delivered directly to a detector or carried through addititional treatment steps. In the latter case, various degrees of dispersion and dwell times can be attained to generate the required mixing and reaction processes.

Above all else, to properly appreciate and apply the utilities of FIA, an understanding of the effects and attainment of variable dispersion is necessary.

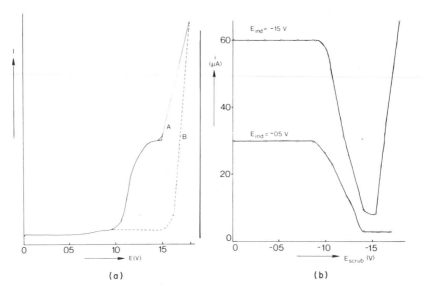

Fig. 15. (a) Voltammogram of eluant recorded at the eluant scrubber (sweep rate of 2 mV/sec.) before (a) and after (b) addition of methanol. (b) Plot of background current in the polarographic detector versus the scrubbing potential for two detector potentials (Eind). [Reprinted with permission from Hanekamp *et al.* (1980).]

Some of the quantitative aspects have already been discussed. However, most FIA systems can be designed quite well and easily using only qualitative models. Therefore, a discussion of the more empirical aspects is in order.

F. Controlling Band Spreading

It should be clear at this point that FIA operates under nonequilibium conditions. The mixing and reaction processes are generally incomplete, but they are reproducible. In addition, these physico–chemical parameters are at odds with an undesirable aspect of axial or longitudinal dispersion, that being excessive sample dilution and band spreading. Therefore, in relatively complex FIA systems, which require that the sample be carried through several steps or where longer reaction times must be accommodated, it is necessary to concomitantly maintain axial dispersion beneath an acceptable upper limit.

Indeed, Vanderslice's group discovered a set of conditions in open bore narrow tubes, where volume dispersion virtually ceases to increase above a given flow rate. Currently, however, a specific formula has not been proposed to limit axial dispersion in manifold designs of various sample volumes, flow rates, tube lengths, and tube bore sizes. As a result, the conditions must be determined

TABLE II

THE BACKGROUND CURRENT AND NOISE AT A DME DETECTOR (WITH COLUMN) AFTER SCRUBBING

Drop size		Small (S)		Medium (M)				Large (L)			
Mode	Drop time (sec)	Baseline (nA)	Noise (nA)	Baseline (nA)	Ratio M:S	Noise (nA)	Ratio M:S	Baseline (nA)	Ratio L:S	Noise (nA)	Ratio L:S
Sampled d.c.	0.5	27.00	2.00	42.50	1.57	3.75	1.88	52.00	1.93	4.50	2.25
	1.0	20.75	1.50	37.00	1.78	3.75	1.77	47.75	2.30	2.50	1.67
N.p.	0.5	25.75	2.50	43.25	1.68	4.50	1.80	57.25	2.22	6.50	2.60
	1.0	21.25	1.50	35.50	1.67	4.00	2.67	48.50	2.28	5.50	3.67
D.p. 25	0.5	0.13	0.28	0.44		0.50	1.79	−0.69		0.64	2.29
	1.0	0.10	0.13	0.09		0.59	2.23	−0.71		0.30	2.31
D.p. 50	0.5	−0.31	0.25	0.85		0.36	1.44	0.34		0.63	2.52
	1.0	0.19	0.10	0.03		0.15	1.50	0.61		0.17	1.70
D.p. 100	0.5	−0.34	0.28	1.48		0.56	2.00	0.60		0.75	2.68
	1.0	−0.50	0.11	0.45		0.10	0.91	+0.15		0.24	2.18
Mean					1.68		1.74		2.18		2.39

empirically. This does not imply that the Vanderslice finding is unimportant or unworkable, only that it requires further development before it can be straightforwardly applied.

Two alternatives have been proposed for limiting axial dispersion as the dwell time is increased. The technically simpler model of Tijssen (1980) involves the use of tightly coiled capillary tubing. In this case, the centrifugal force generated as a result of the coiling induces secondary flow in the radial direction, thus reducing band spreading and carry-over relative to straight open tubes. Unfortunately, the easiest way is often not the best. In spite of its capacity for limiting axial dispersion, capillary tubing is simply not practical to use in systems where particulate matter or precipitatable macromelecules may be present since it will clog too readily.

Fortunately, another approach has been developed, which is only slightly more complex, but which demonstrates significant functional improvements over coiled narrow bore tubing. Van den Berg *et al.* (1980) originally suggested the use of packed bed reactors based on their previous application to post-column derivatization in HPLC. Reijn *et al.* (1981) subsequently introduced the single bead string reactor (SBSR) as a mechanism to limit band spreading that is easily

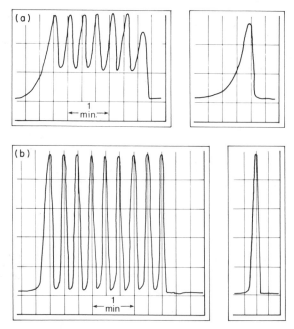

Fig. 16. FIA peaks resulting from injection of a sample followed by flow through (a) an open tube, length 96 cm and (b) a single bead string reactor. [Reprinted with permission from Reijn *et al.* (1981).]

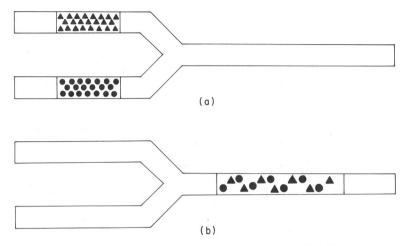

Fig. 17.   Schematic presentation of merging zones principle.

prepared and does not cause a large pressure drop in the system. In a comparative study of their SBSR and coiled open bore tubing they made three important determinations. First, as illustrated in Fig. 16, they found that much less axial dispersion occurred in the SBSR, which allowed higher sample throughput with excellent baseline resolution and an increase in peak height. They also discovered that decreasing the flow rate to increase dwell time did not adversely affect the peak height in the SBSR but did in open coiled tubes. Finally, it was noted that 3 to 4 samples could be accommodated in the SBSR simultaneously without interaction.

G. *Merging Zones*

Zone sampling was discussed earlier in the section on sample pretreatment. This, in fact, was an outgrowth of the principle of merging zones, which was first proposed by the CENA* group (Bergamin *et al.,* 1978). The principle of merging zones is illustrated in Fig. 17. Using this technique, not only the sample, but also the reageant(s) is injected into a carrier stream (a). The two or more zones thereby created are then pumped to a confluence fitting (b) where they merge and partially mix. The combined zones then proceed through a reaction coil to promote further mixing.

Merging zones bestows several advantages on FIA. Among the most obvious is that only precisely the amount of reagent required is incorporated into the system for each analysis, the carrier stream being used for washout. Thus,

*Centro de Energia Nuclear na Agricultura, São Paulo, Brasil.

reagent is significantly conserved, which is particularly important in process monitoring applications to minimize reagent consumption.

In order to appreciate the magnitude of reagent conservation possible with FIA, consider the determination of nitrite/nitrate. As described by Gine *et al.* (1980), the sample, buffer, and color reagent are injected simultaneously. The sample and buffer merge and mix through a coil. This mixture then merges with the color reagent, which has been pumped through a phasing coil to synchronize its arrival at the confluence point with that of the sample/buffer zone. Furthermore, the cadmium reduction column can be switched in and out of the analytical manifold to facilitate alternating determinations of nitrate and nitrite. The amount of color reagent consumed is only 50 μliter per analysis. This means that 3 liters of reagent will support 60,000 analyses. If nitrate and nitrite are each analyzed every 15 min, the monitor could run for over 50 years without the reagent's being replaced, at least in terms of volumetric considerations. Obviously, although this feature is dramatic, practical considerations prevent it from being ultimately realized. However, much smaller reagent containers can thus be incorporated into an FIA monitor as compared with other continuous monitors, thereby contributing to the miniaturization of a self-contained system.

Merging zones can also be used to eliminate several problems that may be encountered when using sample injection alone. For example, the high blank values and negative peaking observed in the analysis of low analyte concentrations could be circumvented by combining the sample and reagent at a confluence point rather than injecting the sample directly into the reagent. The improved mixing thus promoted also increases reaction efficiency, resulting in higher sensitivity. The reason for this is that the reagent immediately reaches the center of the sample zone when a confluence arrangement is used.

H. Titrations

In addition to end point and reaction rate measurements, titrations can also be performed using flow-injection analysis. The ability to automatically titrate an analyte is important due to the precision, specificity, sensitivity, and simplicity of the chemistry. However, until recently the hardware required for automation was complex and expensive.

Titrations depend on a molar equivalent of reagent being consumed in the presence of the analyte, thus the variables normally involved are the concentrations and volumes of the analyte (A) and reagent (B) as expressed by the relationship $C_A V_A = C_B V_B$. A titration can be performed by maintaining any three of the variables at a constant value and varying the fourth. The problem then becomes how to physically accomplish this. As pointed out by Ruzicka in his original description of FIA titrations (Ruzicka *et al.*, 1977), in a flowing stream where substances A and B are continuously mixed as they are in FIA, the flow rates can replace the volumes in the equivalence expression. As a result, either the concentration or the flow rate of one of the substances can be varied to effect a titration.

Since the point of performing a titration is to determine the analyte concentration which varies from sample to sample, C_A is the obvious variable. This is best accomplished by injecting the sample into a carrier stream and delivering it to a mixing chamber which is continuously stirred. The sample is thereby substantially dispersed, and a well-defined concentration gradient is formed.

The actual volume of the analytical line occupied by the sample will be constant from sample to sample. However, the number of analyte molecules per unit volume along the length of the sample zone will obviously vary with the analyte concentration. Therefore, the more concentrated the analyte, the sooner the equivalence point will be reached between the sample and reagent as they are combined. Thus, the portion of the sample zone where the equivalence condition has been fulfilled will vary, and this can be evaluated either colorimetrically or potentiometrically.

In order to resolve these differences, it is necessary to elongate the sample zone as previously described. The analytical readout can then be taken from the peak width as illustrated in Fig. 18. It may be seen that sample throughput in an automated FIA titration system can be limited by the most concentrated unknown, the value of which has to be guessed at anyway. A solution to this problem was introduced by Stewart's group (Stewart and Rosenfeld, 1981)

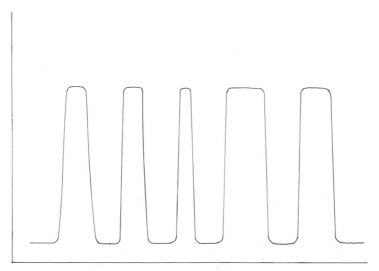

Fig. 18. Typical peak form obtained for FIA titrations.

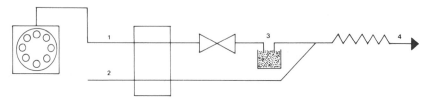

Fig. 19. Analytical schematic of automated FIA titration system: (1) sample, (2) reagent, (3) mixing chamber, (4) to detector.

wherein the automatic sampler is actuated by a feedback loop from the detector, which signals as soon as the post peak baseline has been established for the current sample.

The apparatus required to perform an FIA titration may be as simple as a single reagent stream into which the sample is injected and well dispersed as it is pumped through a length of large bore tubing. However, a far more flexible system, which also extends the linearity, is attained by pumping the sample and reagent streams separately into a mixing chamber as illustrated schematically in Fig. 19.

V. Conclusions

In conclusion, it has been demonstrated that flow-injection analysis is a versatile technique for automating diverse aspects of wet chemical determinations. The salient features include controllable sample dispersion, short response time, high sample throughput, fast startup and shutdown, low reagent consumption, and excellent flow stability. In addition, the various sample pretreatment capabilities described facilitate determinations of a wide range of analytes and concentration ranges in diverse sample matrices. These features combine to form an analytical system that is applicable to both laboratory and on-line monitoring.

References

Basson, W. (1979). *Analyst* **104,** 419.
Bergamin, H., Reis, B. F., Jacintho, A. O., and Zagatto, E. A. G. (1980). *Anal. Chim. Acta* **117,** 81.
Bergamin, H., Zagatto, E. A. G., Krug, F. J., and Reis, B. F. (1978). *Anal. Chim. Acta* **101,** 17.
Gine, M. F., Bergamin, H., Zagatto, E. A. G., and Reis, B. F. (1980). *Anal. Chim. Acta* **114,** 191.
Hanekamp, H. B., Voogt, W. H., Bos, P., and Frei, R. W. (1980). *Anal. Chim. Acta* **118,** 81.
Karlberg, B. (1980). *Anal. Chim. Acta* **118,** 285.
Karlberg, B., and Anfalt, T. (1980). *Pittsburgh Conf.* Paper 82.
Karlberg, B., and Thelander, S. (1978). *Anal. Chim. Acta* **98,** 1.
Kawase, J., and Yamanaka, M. (1979). *Analyst* **104,** 750.

Kawase, J., Nakae, A., and Yamanaka, M. (1979). *Anal. Chem.* **51** (11), 1640.

Nagy, G., Feher, Z., and Pungor, E. (1970). *Anal. Chim. Acta* **52**, 47.

Reijn, J. M., van der Linden, W. E., and Poppe, H. (1981). *Anal. Chim. Acta* **123**, 229.

Reis, B. F., Jacintho, A. O., Mortatti, J., Krug, F. J., Zagatto, E. A. G., Bergamin, H., and Pessenda, L. C. R. (1981). *Anal. Chim. Acta* **123**, 221.

Ruzicka, J., and Hansen, E. H. (1975). *Anal. Chim. Acta* **78**, 145

Ruzicka, J., and Hansen, E. H. (1978). *Anal. Chim. Acta* **99**, 37.

Ruzicka, J., Hansen, E. H., and Mosbaek, H. (1977). *Anal. Chim. Acta* **92**, 235.

Shelly, D., Rossl, T. M., and Warner, I. M. (1982). *Anal. Chem.* **54**, 87.

Stewart, K. K., and Rosenfeld, A. G. (1981). *J. Autom. Chem.* **3**(1), 30.

Stewart, K. K., Beecher, G. R., and Hare, P. E. (1976). *Anal. Biochem.* **70**, 167.

Taylor, G. (1953). *Proc. R. Soc. London, Ser. A* **219**, 186.

Tijssen, R. (1980). *Anal. Chim. Acta* **114**, 71.

van den Berg, J. H. M., Deelder, R. S., and Egberink, H. G. M. (1980). *Anal. Chim. Acta* **114**, 91.

Vanderslice, J. T., Stewart, K. K., Rosenfeld, A. G., and Higgs, D. J. (1981). *Talanta* **28**, 11.

Wallace, V. (1967). *Anal. Biochem.* **20**, 411.

Zagatto, E. A. G., Jacintho, A. O., Pessenda, L. C. R., Krug, F. J., Reis, B. F., and Bergamin, H. (1978). *Anal. Chim. Acta* **101**, 17.

Zagatto, E. A. G., Jacintho, A. O., Mortatti, J., and Bergamin, H. (1980). *Anal. Chim. Acta* **120**, 339.

3

The Monitoring of Cationic Species in a Nuclear Power Plant Using On-Line Atomic Absorption Spectroscopy

GRETCHEN B. GOCKLEY and MICHAEL C. SKRIBA

Westinghouse Research and Development Center
Pittsburgh, Pennsylvania

I. Introduction

A. Background

1. REASONS MONITORING REQUIRED

In recent years interest has mounted in the area of primary chemistry in nuclear power plants, specifically, in relation to purity control of primary-side water. The impurities that are of principal concern are aluminum (Al), calcium (Ca),

69

magnesium (Mg), and silica (SiO_2), because these materials can contribute to the formation of undesirable crud deposits on the nuclear fuel cladding. Therefore, it is desirable to control and/or remove the impurities in the primary water to very low levels [μg/liter or parts per billion (ppb)].

2.   CHARACTERISTICS OF THE PRIMARY COOLANT

Controlling the impurities implies the ability to monitor them. The characteristics of the coolant pose severe constraints on determining a monitoring technique. The background chemical matrix of the primary coolant includes

   (a)   lithium concentration in the range of 0.2–2.2 mg/liter or parts per million (ppm), and
   (b)   boron concentration (a chemical shim to control the activity of the core), which can range from 10–7,000 mg/liter.

Additionally, the radioactivity of the samples necessitates minimizing sample volume and wastes. On-line analysis is desirable to protect the sample from possible contamination by inadequate sampling techniques or sample containers. Another advantage of an on-line system is the ability to obtain real-time data as opposed to the longer turnaround time that is required by grab sampling. No instrument existed that would be directly applicable for on-line primary coolant monitoring of these chemicals at the desired levels.

Of initial importance is the ability of an instrument to monitor the impurities of interest within the chemical constraints of the primary coolant.

The boron concentration in the primary coolant is varied for radioactive flux control purposes. At the beginning of a fuel life cycle the boron will be at a very high concentration to moderate the fuel's activity. The boron concentration, therefore, can be as high as 2000 mg/liter as boron. The boron concentration may also be at the high concentrations when the plant is shut down, either by a controlled, slow addition process or by rapid injections of boron in an emergency situation. At the end of a fuel's life cycle, the boron concentration will be at a very low level in order to optimize the energy production of the fuel. Therefore, the boron concentration can range from a low of 50 ppm to a high of 2000 ppm (as boron).

Another chemical added to the primary coolant is lithium hydroxide. The Purpose of the lithium is to control the pH. The lithium concentration can range from a low of 0.2 mg/liter to a high of 2.2 mg/liter. The concentration is held constant at each plant (at a level in the aforementioned range) and does not vary to a significant extent.

The concentrations of the impurities to be monitored are at extremely low levels, i.e., parts per billion or μg/liter. This is a result of the high purity water required in power plants for corrosion control.

The primary coolant in the nuclear power plant is in direct contact with the fuel rods and, therefore, will become radioactive. In general, the level of radioactivity in the primary coolant will be approximately 80 mrem/hr (millirem per hour) on contact. The radioactivity of the primary coolant can be higher depending upon the condition of the fuel and the fuel's life cycle.

In pressurized water reactors, the primary coolant reaches temperatures of 343.3°C without boiling by maintaining pressures of 2200 psi (pounds per square inch). This condition will cause the coolant to have vastly different chemical properties than it would under manageable sampling conditions.

These chemical and physical properties defined the desired characteristics of the ideal analytical method for primary coolant monitoring. The following characteristics were considered of primary importance:

(a) *Little or no matrix interference* from boron or lithium.

(b) *On-line monitoring capabilities*—to achieve real-time data and to minimize the radioactive waste volume resulting from coolant monitoring.

(c) *Minimum operator attention required*—to minimize operator exposure and time requirements of the monitoring method.

(d) *Operated by relatively untrained personnel*—to minimize the cost to the power plant by not requiring highly specialized analysts.

(e) *Capable of measuring low levels of impurities*—to provide the necessary monitoring of the impurities of interest (calcium, magnesium, and aluminum).

(f) *Reasonable cost*—to minimize the cost to the power plant.

Unfortunately, an ideal analytical method was nonexistent and, therefore, a survey of available methods was undertaken to determine the best method for monitoring low level impurities in the primary coolant.

3. ALTERNATE METHODS OF ANALYSIS

Many different analytical methods were considered, including both direct and indirect methods. The indirect methods investigated were conductivity and sodium measurements. Conductivity measurements are related to the ability of a solution to carry an electric current. The conductivity of the samples is measured by an AC Wheatstone Bridge, potentially the most sensitive, stable, and accurate method available. In the presence of high boron concentrations, it is not practical to use conductivity to measure the very low levels of impurities (μg/liter). Additionally, conductivity will not give specific information in regard to the individual cations, and the "bulk" data of conductivity would have to be related to the individual cation concentrations.

Another indirect method of impurity analysis is by sodium measurement. The most common means of sodium measurement is by specific ion electrode, which is widely used in condensate polishing systems. These systems require sample preconditioning and are sensitive to subparts-per-million levels of lithium. Therefore, the sodium measurement cannot be easily related to the concentrations of calcium, magnesium, and aluminum since lithium is routinely added to the primary coolant.

A variety of direct measurement techniques were investigated. These included ion chromatography, specific ion electrodes, and atomic absorption.

Ion chromatography is a method that utilizes ion exchange resins to separate the ions of interest and then convert them to a form that can be detected by conductivity methods. This method suffers from two major disadvantages as related to this application. One disadvantage is the long analysis time, (18–30 min for calcium and magnesium), and another disadvantage is the length of time required for regeneration of the ion exchange columns between sample analyses.*

Specific ion electrodes for calcium and aluminum were investigated as direct analytical methods. These instruments were found to be incapable of measuring calcium and aluminum in the lower parts-per-billion range (1–10 ppm) due to interferences from the relatively high concentrations of lithium and boric acid.

The final method investigated was atomic absorption. In atomic absorption the sample is heated, either in a flame or a graphite furnace, to sufficiently high temperatures to cause the sample to atomize. The atomized sample will be positioned in the path of light between a hollow cathode lamp source and a detector, generally a photomultiplier tube. The amount of light absorbed can be directly related to the concentration of the element of interest. In flame atomic absorption, the sample is heated in a burner chamber to very high temperatures by use of an oxidant–fuel mixture. The maximum temperature that can be achieved in the burner chamber is dependent upon the gas mixture used. The range of temperatures for various gas mixtures are from 1875°C for air–methane to 2600–2800°C for nitrous oxide–acetylene. The flame method is hindered by three types of interferences; these are termed matrix, chemical, and ionization interferences. In matrix interferences, there may be some constituent in the sample that affects the viscosity of the sample and, therefore, changes the rate at which the sample is introduced to the flame. This can result in either a positive or a negative error in the absorbance readings, because the number of atoms in the light beam will be different. It is, therefore, important to match the standards to the samples as precisely as possible.

Chemical interferences result when the sample contains a constituent that will form a thermally stable compound with the analyte that will not completely decompose in the flame. Two solutions to this interference are most commonly used. One solution is to use a hotter flame, one which will decompose the compound that is formed. The second solution is to add an excess of another

*See Chapter 1 on ion chromatography.

element that will also form a thermally stable compound with the interfering constituent.

Ionization interferences occur when the flame used results in sufficient thermal energy to totally separate the electron from the atom, creating an ion. As a result, there are less ground-state atoms available for absorption, and the absorbance signal is reduced. This interference can be eliminated by adding an excess of another easily ionizable element, which will suppress the ionization of the element of interest.

For this particular application, flame atomic absorption was not chosen for the following reasons:

(a) The detection limit for aluminum was not sufficiently low.

(b) The boron concentration caused chemical interference problems.

(c) The unattended operation of an open flame in the operating portion of a nuclear power plant is unacceptable (Barnard, 1979; Beaty, 1978; Cragin and Herron, 1973; Slavin, 1975).

The instrumental method that most closely met the criterion of an ideal method was the furnace (or flameless) atomic absorption spectrophotometer. (Table I summarizes the methods that were investigated and their respective advantages and disadvantages.)

In graphite furnace atomic absorption, the sample is introduced (by pipetting) into the furnace (Hageman *et al.*, 1979; Hosking *et al.*, 1979; Krasowski and Copeland, 1979). The graphite furnace is held between two large graphite rings, and the furnace in between acts as a resistor in an electrical circuit. A voltage is applied to the tube, and the amount of current flow controls the temperature of the tube. The tube can reach temperatures of 3000°C and is water-cooled to return the tube to room temperature in preparation for the next sample.

There are two gas purges in a furnace system. The external purge shrouds the furnace from the oxidizing atomosphere. The internal purge enters the furnace from the ends and leaves by way of the center hole. This purge operates only when the furnace is heating up and aids in eliminating matrix interferents, controls furnace sensitivity, and reduces background absorption problems.

The furnace is heated in a particular sequence of drying, charring, and atomizing (Hagemen *et al.*, 1979; Hosking *et al.*, 1979; Krasowski and Copeland, 1979; Slavin and Manning, 1979). The drying step is performed at approximately 100°C to remove any water from the sample. Next, the sample is charred to remove as much of the matrix as possible. Finally, the furnace temperature is rapidly raised to a temperature sufficiently high to atomize the analyte. The amount of light absorbed is detected and recorded, then the furnace is cooled. Figure 1 illustrates the furnace program. The increase in temperature can be accomplished in a rapid step-wise fashion, or at a slow, controlled rate, or

TABLE I

Comparison of Analytic Methods for Trace Measurements
of Calcium, Magnesium, and Aluminum

Method	Principle of measurement	Advantages	Disadvantages
Indirect			
Conductivity	Wheatstone Bridge	Simple, easy, inexpensive, accurate, fully automated	Indirect, not practical in presence of boric acid
Sodium	Specific ion electrode	Accurate, automated, on-line. Continuous measurement, commercially used	Indirect, sensitive to lithium and use of glassware
Direct			
Ion chromatography	Ion exchange and conductivity	Accurate, automated, on-line, good detection limits	Sole distributor, long analysis time, incapable of Al measurement
Specific ion electrode	Specific ion electrode	Accurate, automated, on-line, continuous, no interferences by boron or lithium	Requires verification for sub-ppb for Al and Ca; no method for Mg
Atomic absorption			
Flame	Light absorption	Accurate, automated	One flame, insufficient detection limit, boron and lithium interference
Furnace	Light absorption	Accurate, automated on-line continuous method, little interference from boron or lithium, low sample volume, short analysis time, computer interfaceable	Expensive high operating costs, life of furnace affected by boron concentration

"ramp" (as shown in Fig. 1). The manner in which the temperature is increased depends upon the step in the program, the element of interest, and the sample matrix. The furnace program is obviously highly variable and must be adjusted for each element to achieve the best results.

The instrumental method had some excellent advantages for this particular application.

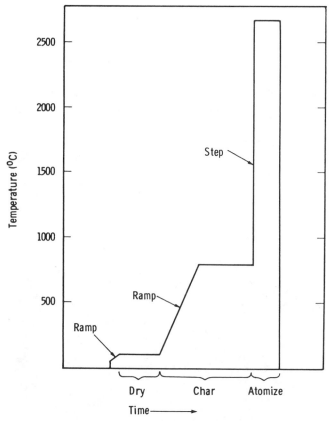

Fig. 1.   Diagram of graphite furnace program.

(a)   Ultralow concentrations [sub-μg/liter or subparts per billion (ppb)] of the elements of interest can be analyzed by flameless atomic absorption.

(b)   Atomic absorption is a multielement method of analysis (Alcock, 1979; L'vov, 1978; Manning, 1978) and, therefore, a single instrument can accomplish all three necessary analyses.

(c)   This method is a direct method of analysis and, therefore, no skepticism would be associated with the data.

(d)   The analysis time is short (i.e. 2 min), which can allow fast turnaround time and a large number of samples to be done in a short period of time.

(e)   The required sample volume is small and, therefore, the method is suited to the radioactivity of the sample.

(f)   Instrumentation for atomic absorption was available with computer controls and interfacing. This results in automatic operation of the instrument, which will limit the amount of time necessary for operator attention.

B.   Description of the Atomic Absorption System

Atomic absorption techniques have been significantly improved with the addition of automatic sampling systems by upgrading the precision and the efficiency of the analytical method as well as increasing the throughput of the laboratory. These systems offer the advantage of a mechanical means of withdrawing a sample and dispensing that sample into the graphite tube without operator intervention. This reduces the possibility of sample contamination and carryover frequently occurring in manual injections.

For this particular instrument, two small piston pumps (a sample pump and a flushing pump) are mounted underneath a circular sample tray (See Fig. 2). The

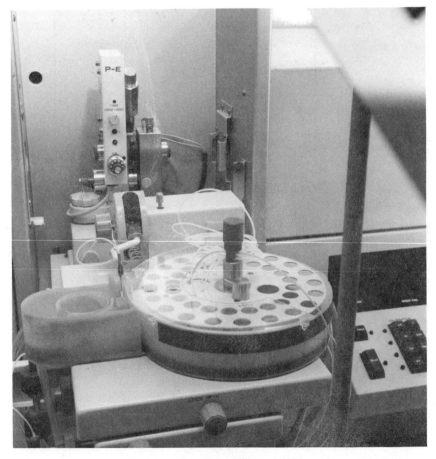

Fig. 2.   Automatic sampling assembly.

pumps are connected to a sample arm which rotates between the sample tray and the furnace part in the following manner:

(a)   The capillary tip of the sample arm is positioned in an overflowing cup of rinse water (to the side of the sample tray) while in the rest position to constantly clean the tip.

(b)   The flushing pump (connected to the blank or flush solution) cycles to rinse the capillary tip internally and externally. This fills the sample arm with blank (or flush) solution.

(c)   The sample arm is positioned directly above the overflowing cup and a 25-μliter air bubble is drawn into the arm to separate the sample and the flushing liquid.

(d)   The sample arm then indexes to the programmed sample cup on the sample tray. The sample arm is lowered into the cup, and the preselected sample volume is drawn into the arm.

(e)   The sample arm then swings over the the furnace, positioning the capillary tip in the introduction hole in the graphite tube.

(f)   The liquid is dispensed into the graphite tube by the pump.

(g)   The sample arm returns to the rest position in the overflowing cup.

(h)   The furnace program then begins. The sample tray will rotate to the next assigned position (or return to previous sample cup for multiple analysis on the same sample), and the sequence begins again.

The sample trays contain 40 sample cups of 2-mliter capacity. The sample tray is covered to reduce possible evaporation of the sample and to prevent sample contamination by air-borne materials.

The system controller for the AA system controls both the furnace programmer and the spectrophotometer. The controller instructs the spectrophotometer when to zero or standardize, and the furnace programmer when to execute its program. In addition, the controller may call up new analytical programs from the internal locations of the spectrophotometer and the furnace systems. The furnace programmer will signal the spectrophotometer to read and will signal the controller when it has completed its program so that a new sampling cycle may begin.

Several analytical methods are easily accomplished with the controller. The first method involves automatic calibration of the spectrophotometer and sequential analysis of up to 35 samples. Another method available with the controller is an "automatic" standard additions technique. This is useful when it is difficult to prepare standards with a similar matrix. The third method available is an automatic method-of-additions technique in which the standard is added directly to the graphite tube.

All of these methods offer the operator the advantage of automatic analyses.

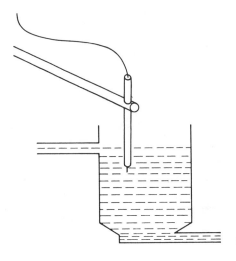

Fig. 3. Schematic of flow-through sample cup.

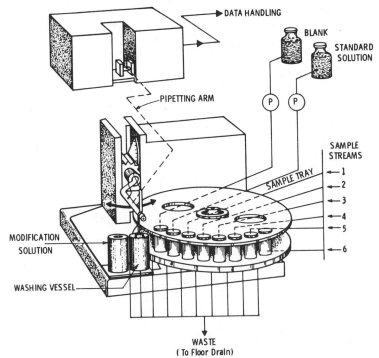

Fig. 4. Schematic of sampling system.

Fig. 5. AA—installed in power plant.

However, the operator must program the instrument and fill the sample trays. For this application it was necessary to minimize contamination and operator requirements; therefore, an on-line system was required. One manufacturer offers a modified version of their controller, which performs fully automated monitoring of a flowing stream. The major advantage offered by this system is that it had previously been used successfully in a monitoring function in Germany. The sample tray for on-line automatic operation of the system contains 8 small overflowing sample cups. In these sample cups, shown in Fig. 3, the flowing sample stream is introduced at the bottom and allowed to overflow at a specific level near the top (with the overflow being sent to a drain or waste bottle). Therefore, the sample cup has a continuously fresh sample for analysis. Two of these sample cups are for blank and standard solutions. Therefore, the autosampler performs automatic zero setting and concentration calibration of the spectrophotometer and automatic analysis of up to six sample streams.

During operation, the sample streams are constantly overflowing (at a rate from 10 μliter/min to 40 μliter/min), and the standard and blank are pumped at intermittent rates, flushing the system totally (including the cups) during the flush cycle. The blank and standard solutions are pumped from 1-gal jugs, placed on weight switches that are connected to the programmer. In the event that the level in either of these reservoirs should become too low, the programmer interrupts the program, holding all programs in the memory, until the reservoir is refilled and the system restarted. Figure 4 is the schematic representation of the controller.

The AA Spectrophotometer system is a sophisticated laboratory instrument, and several adaptations were necessary to integrate the AA into the overall system and permit it to operate in the auxiliary building of the power plant. Figure 5 shows the instrument as installed in the power plant. The objectives to be achieved were (1) to show that analytical techniques were available for routine analyses of parts per billion of the contaminants of interest, and (2) that the analyses could be done on-line in an operating power plant. The adaptations required for this application fell into two categories: adaptations for the plant

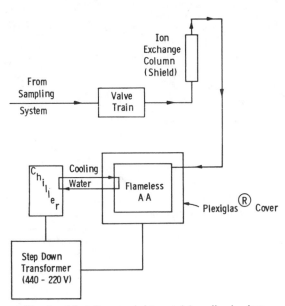

Fig. 6.   Adaptations needed to put AA on line in plant.

environment and modifications required by laboratory experience. The adaptations required for operation in a plant environment included a valve train, a chiller, and a plexiglas cover. An ion exchange cartridge and a transformer were added owing to laboratory experience. These adaptations are shown schematically in Fig. 6.

II. Adaptations for the Plant Environment

A. Valve Train

The sample stream to be analyzed was a bleed line from the existing sample system of the primary coolant system. A valve train was needed to prepare the stream for the AA system. The valve train's purpose was to meter the flow, protect the downstream components from overpressure, and remove relatively large suspended solids. Figure 7 is a schematic of the valve train, and Figure 8 is the valve train as installed in the plant. The first valve is an isolation valve (not shown in Fig.7). This valve was opened to allow flow through the valve train and closed when the system was not in use. The next valve is a pressure-reducing valve to reduce the pressure from a possible maximum of 2200 psi and provide a constant head supply to the in-line filter (which removes suspended solids) and the micrometering valve which provides precise control of the sample flow rate. This portion of the valve train is protected by a pressure relief valve which opens to the drain. After the metering valve, the sample flows to the ion exchange cartridge, which has a design pressure limit of 5 psi. Therefore, it was necessary that this portion of the system be protected by a pressure relief valve similar to the previous valve. The valve train was of all stainless steel construction and functioned very well. Only occasional monitoring was necessary with some minor adjustments when the primary coolant system exhibited wide variations in pressure.

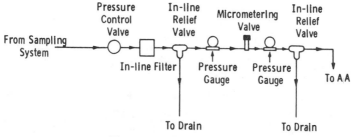

Fig. 7.   Valve train schematic.

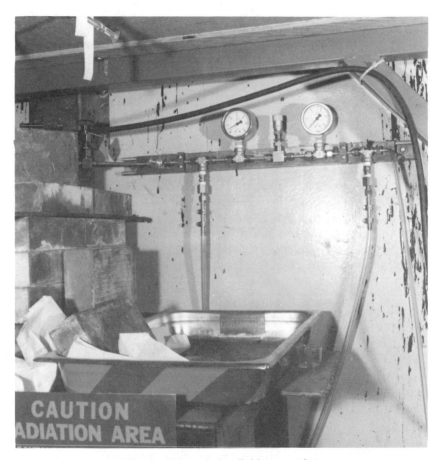

Fig. 8. Valve train installed in power plant.

B. *Chiller*

The AA unit requires a cooling water flow to the optics and the furnace holder to cool these components in between analyses. In the laboratory, this is normally provided by tap water, but in the restricted area of the power plant any waste water must be treated as radioactive and, therefore, is carefully controlled. To provide the necessary cooling water and to keep the water use to a minimum, a small portable chiller was used with recirculating cooling water. A small 5000-BTU/hr add-on system was purchased and was more than adequate for this system.

C. Plexiglas Cover

In a plant installation one of the obvious concerns is dust and dirt protection. During operation the AA system offered some built-in protection against airborne contamination. The gas sweep through the furnace prevents dust from settling in the furnace and the on-line, closed pump system offers protection from contamination introduced by manual injection. However, in the area immediately adjacent to the instrument, bags of cement were stacked for interim storage, and construction activities requiring core drilling of the concrete wall were conducted. Therefore, to protect the instrument from these significant dust problems, the entire system was covered by a simple clear plastic cover with sliding doors in the front. Provisions were made to ventilate the box and to provide filtered make-up air; however, this was never needed.

The protective cover offered another benefit. The instrument had been installed in a relatively high traffic area. The locked doors provided protection against unintentional and/or improper operation of the instrument. This was particularly helpful when the instrument was operating unattended, since it could be termed somewhat of an "attractive nuisance" with its rotating sampling arm and lighted panels.

III. Modifications Required by Laboratory Experience

A. Ion Exchange Cartridge

Initial tests done in the laboratory prior to the instrument's being installed in the plant indicated that a means of removing boric acid from the stream was needed, and an isolation transformer for electrical protection was necessary also.

Boric acid is used in the nuclear reactor as a chemical shim (as described earlier). The boric acid concentration ranges from 100 ppm to levels as high as 1500 ppm, with the very high levels encountered during shutdown operations. Since these tests were to be conducted during a routine shutdown, the effect of boric acid on the analysis of calcium, magnesium, and aluminum was important. During this laboratory testing, it was discovered that the boric acid caused absorbance signal suppression. This effect is shown in Fig. 9. The expected impurity levels were less than 10 ppb, and therefore at the expected boric acid concentrations the absorbance signal was so low as to be unreadable. In addition the boric acid gradually attacks the graphite tube, causing it to deteriorate at a much higher rate than desirable. Some means of circumventing these problems was needed.

The boric acid can be removed in several ways. One standard method is the

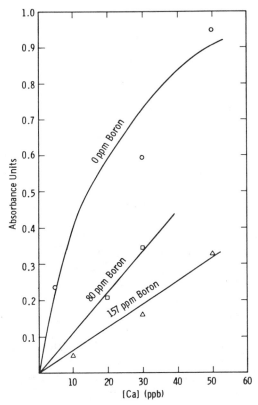

Fig. 9.   Effect of boron concentration on calcium absorbance signal. ○, 0 ppm boron; □, 80 ppm boron; △, 157 ppm boron. Sample size = 10 μliter.

use of methanol. The addition of methanol forms volatile methyl borates, which vaporize in the heatup stage, thereby removing the interfering boron. This has been quite successful in the laboratory. However, to accomplish this in the plant a reservoir of the methanol would have to be placed next to the sample tray. Since the methanol is highly flammable, this procedure is not acceptable for the operational areas of a nuclear power plant. A companion method to this was proposed immediately prior to the shutdown of the plant. In this method the furnace temperature is increased with a slow ramp and held at a temperature sufficiently high to drive off the volatile boric acid and leave the nonvolatile elements of interest behind. Unfortunately, the limitations of time prohibited worthwhile investigations of this method.

Previous work had indicated that boric acid could be stripped out of the primary coolant stream without affecting the cation concentrations. This work had been done using ion exchange resin, in particular strong base, nuclear grade anion resin in the hydroxide form. For this application a clear, inert, sealed plastic cartridge, filled with 0.085 ft^3 of resin was used. This cartridge could be changed quite easily since the end fitting was of the hose nipple configuration. The cartridge was shielded owing to the fact that when it was filled with primary coolant there was a measurable amount of radiation on contact. The flow entered from the bottom of the cartridge and exited from the top. This prevented possible channeling through the cartridge and ensured sufficient contact time for the removal of boric acid.

In the eight months of intermittent analyses, only two cartridges were used for these tests. One broke through after approximately 600 hours of operation at a low boric acid concentration, the other broke through after approximately 100 hours of service at high boric acid concentration.

B. Transformer

During laboratory tests, it was discovered that electrical line noise affected the electronics of the AA system. Large loads tripping on the laboratory circuit occasionally gave spurious readings. To overcome this problem an isolation transformer was used during most of the laboratory shakedown. In the area in which the instrument was installed in the plant only 440–V power lines were available. Therefore a 440/220 step-down transformer was necessary, and this served also as an isolation transformer to protect the electronics. In other applications, where a 220-V supply is available, an isolation transformer is seriously recommended.

IV. System Performance

The AA system functioned exceedingly well in the plant environment and was capable of performing the required low level analyses in nonlaboratory conditions. Owing to very tight time constraints, most of the operating time of the instrument occurred in the plant, rather than the laboratory. However, some initial work that was accomplished in both the laboratory and the plant needs to be described in terms of its relative importance in this application.

A. System Calibration

The effect of sample size was investigated to determine the optimum sample size required for the analysis of calcium, magnesium, and aluminum in the primary coolant and the effect of the puddle size in the furnace. These data are

presented graphically in Fig. 10. As would be expected, the graph illustrates that
as the sample size increases, the absorbance signal increases in direct proportion
until nonlinear absorbance numbers are reached (the concentration of the sample
solution was held constant). This is owing to the fact that, as the sample size is
increased, a greater number of atoms are placed in the graphite tube and, therefore, are atomized during analysis, creating the increase in the absorbance signal.
The trade-offs to be considered for this application involved using a sufficient
sample volume that would yield a satisfactory absorbance signal and yet be as
small as can reasonably be pipetted with the automatic system to minimize
radioactive contamination of the AA system.

The effect of sample puddle size within the graphite furnace is of concern
because of the possibility that a "large" puddle will exceed the isothermal zone
in the furnace. This would result in only partial atomization of atoms present at
the time the spectrophotometer is programmed to "read" the absorbance. Therefore, an inaccurate absorbance reading would result. These considerations resulted in a choice of 10-μliter sample size for calcium and magnesium analysis
and 20-μliter sample size for aluminum analysis. The larger size for the aluminum is warranted owing to its minimal absorbance signal response.

As discussed earlier in this chapter, the boric acid interference was quite

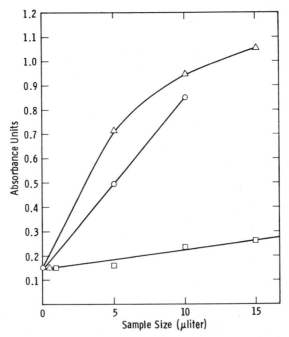

Fig. 10.   Effect of increasing sample size on absorbance signals.

pronounced for the primary coolant samples. The initial problems encountered with the impurity monitoring were a result of the boric acid interference, and the solution was intended to precondition the sample by removing the boric acid with an ion exchange resin.

Although removing the boric acid ameliorated the major problem in the primary coolant monitoring, the initial tests indicated that the standards prepared for this work did not sufficiently match the matrix of the primary coolant. The major constituents, boric acid and lithium hydroxide, were investigated and determined not to be the appropriate matrix interferents. To circumvent this problem, the following calibration method was used:

(a) Standard solution samples that contained 0, 1, 2, or 5 ppb of the elements of interest were analyzed. Generally, the two standards used were 0 and 5 ppb.

(b) A curve was plotted by subtracting the zero (or blank) reading from the 5 ppb standard reading and plotting the straight line between the resultant difference and the origin.

(c) The primary coolant sample is then analyzed, and an absorbance reading for the empty furnace is also obtained.

(d) The difference between the sample absorbance reading and the empty furnace reading is then plotted on the curve drawn from the data (Boutron and Martin, 1979; Christensen, 1972; Isaaq and Young, 1977; Ragone and Finelli, 1971; Siemer, 1978).

This methodology overcomes problems with the matrix interferences quite successfully. Table II is a simple description of this data handling process.

The method of standard additions was also investigated as a means of overcoming matrix interference problems. With this system, the operation can be handled automatically simply by providing the proper standard and programming the controller for this method. The method involves adding known amounts of the element of interest to a constant amount of unknown sample. In addition, the sample alone is analyzed. The curve is drawn, as shown in Fig. 11, with the portion of the x axis to the right of the y axis labeled as the amount of the element added to the sample and the portion to the left labeled as the amount of the element in the unknown sample. Detailed discussions of the method are available in most atomic absorption references. This process was successful, but time-consuming, for an on-line process and as such should only be considered for intermittent usage as a means of checking the data.

One of the objectives of this program was to investigate the feasibility of on-line impurity monitoring to overcome the inherent problems associated with grab sampling. However, after the instrument had been installed in the plant, some interest was generated in improving the grab-sampling process. Attempts were made to monitor standards, and samples of the ultrapure water were used for the

TABLE II

Description of Data Handling

Sample description	Absolute absorbance reading	Corrected absorbance reading	Concentration
Standard	0.1894	0.1000	5 ppb
Blank	0.0894	0.0000	0
Sample	0.0526	0.0500	2.5
Furnace	0.0026	0.0000	0

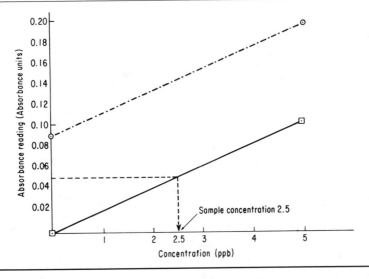

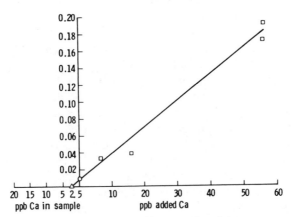

Fig. 11. Standard addition curve for calcium.

standards over a period of several days. In addition, the samples were split; one portion of each was stored in an untreated bottle, and the other was stored in a pretreated bottle. The results of these data are doubtful, and no direct conclusions could be set forth. However, the general indication is that grab sampling and possible bottle contamination may affect the analysis of samples and, therefore, on-line monitoring at ultra-low levels is highly advisable (Christensen, 1972; Isaaq and Yound, 1977; Ragone and Finelli, 1971).

In the plant, the AA system operated quite well. The conditions were not immediately conducive to ultralow level analyses; however, with minor adjustments for the more rigorous environment, the system could be handled quite easily.

One of the initial drawbacks to operating in a power plant was the laboratory conditions available for support work. To prepare the ultralow level standards necessary for these tests, a relatively clean area, ultraclean glassware and measuring apparatus, and high purity standards are required. In general, plant laboratories do not have these facilities available and, therefore, an area of the laboratory was cleaned for the express use of the personnel involved in this program. Additionally, some glassware, pippettes, and high purity standards were provided for the program by an independent laboratory (Hosking *et al.,* 1979; Krasowski and Copeland, 1979; Slavin and Manning, 1979). Extreme care had to be taken to prevent contamination of the samples, and when such was provided, the procedure resulted in excellent standards. Periodically, through the testing program the standards were verified by sending samples to outside laboratories. The results were generally such that the standards could be considered quite accurate.

A primary concern in on-line monitoring should be the placement of the instrument. Long expanses of sample line only result in making one more skeptical of the final data. Therefore, the instrument should be located as near as possible to the sample source. An added complication in this instance was the radioactivity of the sample. All sample lines would have required some form of shielding and would have involved special monitoring for leak protection (Christensen, 1972). In addition, the holdup time in the sample lines can create problems in relating the monitoring data to the plant operations that are to be monitored. For this application, the instrument was installed against the wall opposite the existing plant-sampling system. This placed the instrument in a noisy, high-traffic area; however, no problems were encountered owing to the position of the instrument.

By using the existing sampling system, some problems were encountered which resulted from the lack of control and communication between the plant operating personnel and the test program personnel. One problem was inconsistency of the feed to the AA sampling line. During normal plant operations, the sample point for the AA may be changed by the plant operators from the reactor hot leg sample (primary coolant leaving the reactor and entering the steam generators) to the pressurizer vapor space. This is because venting of the pressurizer is accomplished through the sample line to the volume control tank to reduce the hydrogen concentration in the primary coolant. However, the sprays into the pressurizer vapor space are on at this time, producing confidence that this sample is always representative of the primary coolant. This procedure, though, was signalled at the instrument by a drastic reduction in the sample flow, and generally, the instrument operator is not informed of the change. However, when possible, a hot leg sample was used for data collection.

As the shutdown operations proceed, the temperature and pressure of the primary coolant are gradually reduced. Since the valves in the valve train were set for operating system pressure, it became necessary to adjust these valve settings so as to continue to produce the necessary flow through the system.

B. Instrument Operational Experience

The major portion of instrument service was incurred in the power plant, and therefore, the instrument problems that occurred were not apparent in the short time the instrument was used in the laboratory. As a result, although the instrument problems encountered did occur in the plant, there is no direct evidence that the problems are a result of plant installation. The problem that reoccurred with the greatest frequency was the spectrophotometer reading and printing without a command from the furnace control unit. This resulted in the print-out's being a list of numbers ranging from 0.000 (when the furnace was not yet hot) to 0.1 (when the furnace was at some undetermined point in its program). When this occurred, it was impossible to determine if any of the printed data was the correct absorbance reading. No cause for this was ever discovered; however, it was possible to correct it by shutting the instrument down and restarting it after approximately 30 min. This resulted in the instrument's functioning properly about 75% of the time after being restarted.

Another reoccurring instrument problem is the loss of memory of the wavelength controlling motor. This caused the energy to be zero and, therefore, the absorbance readings would also be zero. Reprogramming the spectrophotometer circumvented this problem but, again, the cause was not ascertained.

The most difficult problem with the spectrophotometer was its occasional tendency to invent its own programs. During plant operation, the instrument was programmed for three separate sets of conditions, with a command to repeat the first program after completion of the third. However, rather than repeating the first program, the instrument would assign wavelengths, energy levels, etc., to a fourth program and begin analysis. Reprogramming usually solved this problem, but it occasionally would reoccur.

Another instrument problem involved tube alignment. Although the tube was properly aligned, some fine adjustments can be made to eliminate all but a small portion of the empty furnace absorbance signal. This can be a time-consuming process, and since the tubes were changed on a frequent basis, the decision was made to do only the necessary tube aligning.

On occasion, the automatic sampling arm would enter the furnace at an improper time, prior to the completion of the furnace cooling cycle. Since the furnace was still hot, the tip of the sample arm would then melt. The remedy for this was simply to trim the end of the tubing. This generally occurred during manually operated tests.

Time constraints were quite tight once the instrument had been installed in the plant and, therefore, servicing of the instrument for these problems was not advisable since they could be circumvented without involving a significant amount of time. As previously mentioned, the instrument was not positioned in a way conducive to servicing. The height and placement made access to the back of the units difficult. It was not until after the program had been completed and the plant was shut down that adequate time and space were available for servicing. Some minor problems were discovered once the instrument had been serviced; however, the instrument has not been used for a sufficient period of time to determine if complete solutions were found for all of these problems.

The time required to solve the instrumental problems during the on-line monitoring peroid was minimal owing to the preprogrammed magnetic cards. The instrument was turned off for approximately 1 min then restarted, programmed with the cards, and then the operations were initiated. The entire process required a maximum of 5 min.

The time required for the training of personnel, installation, and operation of this AA system is relatively short considering the sophistication of the technique. The instrument and associated program controllers have been designed for the use of the novice as well as the experienced spectroscopist. The engineer, chemist, or spectroscopist, with varying degrees of instrumental experience, can be trained to program the instrument, analyze the data, and adjust existing programs in a few weeks. Technicians, not experienced with this particular instrumental technique, can be trained to operate an already programmed system in a few days, and an experienced technician could be trained to operate, program, and analyze the data in a few weeks. Therefore, the instrument is ideal for plant operations. Installation of the instrumental system can be accomplished in one or two days, provided that the area is prepared for the instrument. Other time may be involved in order to provide the sample stream, instrumental space, and adequate protection.

Operating time is quite minimal; to start a programmed instrument approximately 5 min is required. However, to reprogram with magnetics cards requires only another 2 or 3 min and to reprogram using the keyboard may require approximately 15 to 20 min. This is totally feasible for most plant operating conditions.

V.   Test Conclusions

Although the technique of flameless atomic absorption spectroscopy is a highly sensitive method for low level analysis, it has been shown that with proper installation, it is possible to conduct on-line analysis in a plant environment without its being necessary to use a highly trained spectroscopist. Some sample pretreatment and instrument protection may be required to create the most conducive environment for instrumental operations. However, as a result of this work, the obvious conclusion is that any analysis that can be done off-line using grab samples could be accomplished on-line with this system. The greatest advantages are that the analysis can be done on-line to monitor the impurities in a sample stream with little danger of sample contamination and that it provides information in real time rather than with delays because of laboratory analyses, and all the while providing a minimal amount of operator exposure to the sample stream. Another advantage is that the instrument provides information on each of the elements individually rather than a lumped parameter, as in the case of conductivity. However, it should be remembered that this instrument is not sufficiently developed to operate in a control function. As a complete system, the spectrophotometer and associated components can be used as an alarm function to indicate that further examination of the data may be required to determine fully the necessary actions.

ACKNOWLEDGMENTS

The authors wish to acknowledge J. M. Otte for his dedication and hard work, which contributed greatly to the success of this project. Additionally, M. L. Theodore, C. L. Page, and Z. L. Kardes, were very helpful providing guidance and technical assistance in the use of analytical techniques and flameless AA methods.

References

Alcock, N. W. (1979). *At. Absorpt. Newsl.* **18**, 37.
Barnard, T. W. (1979). *Anal. Chem.* **51**(12), 1172.
Beaty, R. D. (1978). "Concepts, Instrumentation and Techniques in Atomic Absorption Spectrophotometry." Perkin-Elmer Corp., Norwalk, Connecticut.
Boutron, C., and Martin, S. (1979). *Anal. Chem.* **51**(1), 140.
Christensen, S. (1972) *At. Absorpt. Newsl.* **11**,(2), 51.
Cragin, J. H., and Herron, M. M. (1973). *At. Absorpt. Newsl.* **12**(2), 37.
Hagemen, L., Mubarak, A., and Woodriff, R. (1979). *Appl. Spectrosc.* **33**(3), 226.
Hosking, J. W., Oliver, K. R., and Sturman, B. T. (1979) *Anal Chem.* **51**(2), 307.
Issag, H. J., and Young, R. M. (1977). *Appl. Spectrosc.* **31**(2), 171.
Krasowski, J. A. and Copeland, T. R. (1979). *Anal. Chem.* **51**(11), 1849.
L'vov, B. V. (1978). *Spectrochim. Acta, Part B.* **33B**, 153.
Manning, D. C. (1978). *At. Absorpt. Newsl.* **17**(51), 107.

Price, W. J. (1979). "Spectrochemical Analysis by Atomic Absorption." Heyden, London.
Ragone, S. E., and Finelli, R. (1971). *At. Absorpt. Newsl.* **10**(6), 126.
Rawa, J. A., and Henn, E. L. (1979). *Anal. Chem.* **51**(3), 452.
Siemer, D. D. (1978). *Environ. Sci. Technol.* **12**(5), 539.
Slavin, M (1978). "Atomic Absorption Spectroscopy." Wiley, New York.
Slavin, W., and Manning, D. C. (1979). *Anal. Chem.* **51**(2), 261.

4

The Automation of Laboratory Gas Chromatographs for On-Line Process Monitoring and Analysis

JOSEPH P. HACKETT and GERST A. GIBBON

Pittsburgh Energy Technology Center
U.S. Department of Energy
Pittsburgh, Pennsylvania

I. Introduction

Gas chromatography, since its introduction in 1952 (James and Martin, 1952), has proven to be the most powerful analytical technique available to the chemist for the analysis of gases and complex mixtures of volatile substances. The enormous scope of the technique is illustrated by the large number of publications appearing annually (Varian Associates, 1977). Instrument manufacturing companies, realizing the potential market, were quick to produce commercial equipment and undertook investigations leading to the development of gas chromatographs with a high degree of sophistication and reliability. The speed with which chromatographic separations can be obtained and repeated, and the ease with which detector signals can be registered and repeated has led to the employment of the technique as a means of process control. Process control chromatographs are extremely effective for their design function, namely, monitoring small changes in reasonably constant analytical matrices. This high specificity becomes a disadvantage, however, when the process gas stream being monitored is undergoing rapid and wide changes in chemical composition, as is not uncommon during process development. A possible sequence of a tail gas composition during an experimental run of a coal liquefaction or gasification process development unit might include pure nitrogen during initial purge, unreacted feed gas during startup, and a variety of product compositions during the parameter variation phase of the experiment. One rather common approach to the problem of monitoring the gases, generated during the operation of such a process development unit, is the taking of "grab" gas samples for laboratory analysis. This approach has several disadvantages: it is labor intensive; it requires the handling and storage of sampling containers; and it is slow relative to the time scale of an on-line analysis.

The Process Monitoring and Analysis Branch of the Pittsburgh Energy Technology Center, U.S. Department of Energy, has developed an alternative to the "grab" sampling approach that eliminates many of its disadvantages and consequently improves the quality of the gas analysis. Automated research-quality gas chromatographs are attached directly to the coal gasification or liquefaction equipment. The degree of automation of the on-line chromatographic system is determined by the analyst and may include fully automatic stream selection, sample injection, recalibration of the chromatograph, and the reporting of analytical results. An automated on-line monitoring system requires virtually no operator attention during routine operation.

In addition to eliminating the sampling errors and easing manpower needs and time delays, the on-line gas chromatograph offers several additional benefits: the chromatograph can be optimized to the specific process; calibration standards can be better matched to the process stream; and the computerized data system can be interfaced directly to the process control computer for data storage and further analysis.

This chapter provides (1) a review of the laboratory chromatographic technique, presently used at PETC, to analyze "grab" gas samples from process development units, (2) a detailed explanation of what was required to adapt the laboratory technique for automated, on-line analysis, and (3) several applications demonstrating the concept. The chapter is written as a guide to the analyst who wishes to take a proven laboratory chromatographic separation procedure and move it to an on-line station in an experimental or research area.

II. Laboratory Analysis with Gas Chromatography

The Process Monitoring and Analysis Branch of the Pittsburgh Energy Technology Center is responsible for the quantitative analysis of gaseous mixtures submitted by those involved in coal conversion and utilization research at the center. The branch has three Hewlett Packard 5730 model gas chromatographs dedicated to handling the routine gas analysis sample workload. The instruments are set up to yield quantitative data on those gases listed in Table I, column 1, along with their estimated concentration in each routine sample. Five gases (H_2, N_2, CH_4, CO, and CO_2) either singly or in some combination usually constitute from 90–99 vol % of each routine sample. The hydrocarbons (up to 4 carbons) H_2S, and H_2O (vapor) make up the other 1–10 vol %.

The three chromatographs used for the routine gas analysis are equipped with dual thermal conductivity detectors, gas sampling valves, carrier gas inlets with purification trains, and dual pen recorders. The columns used are (1) 5 Å molecular sieve–60/80 mesh, ⅛-in. o.d. × 10-ft stainless steel with argon as the carrier gas and (2) Porapak "R"–80/100 mesh, ⅛-in. o.d. × 12-ft stainless steel

TABLE I

CONCENTRATION RANGE OF CALIBRATION STANDARDS FOR LABORATORY CHROMATOGRAPHS

Component of interest	Volume percent[a]			
	Estimated concentration range	Laboratory standard no. 1	Laboratory standard no. 2	Laboratory standard no. 3
H_2	0.1–100	24.32 ± 0.24		
N_2	0.5–100	25.63 ± 0.10	Balance	Balance
CH_4	0.5–100	24.84 ± 0.10		
CO	0.5– 35	25.38 ± 0.10		
CO_2	0.1– 35		12.02 ± 0.07	
C_2H_4	0.1–2.0		1.24 ± 0.01	
C_2H_6	0.1–5.0		1.26 ± 0.01	
C_3H_6	0.1–2.0		1.24 ± 0.01	
C_3H_8	0.1–3.0		1.27 ± 0.01	
i-C_4H_{10}	0.1–2.0		1.27 ± 0.01	
1-C_4H_8	0.1–2.0		1.26 ± 0.01	
n-C_4H_{10}	0.1–2.0		1.24 ± 0.01	
He	0.1–5.0			2.05 ± 0.03
H_2S	0.1–3.0			1.82 ± 0.02

[a] Estimated uncertainty at the 95% confidence level.

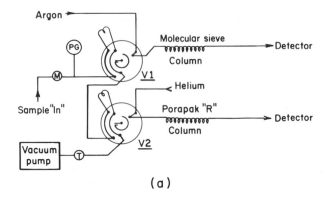

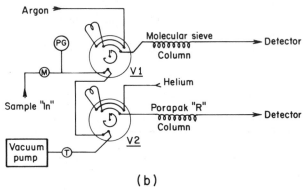

Fig. 1. Sample introduction (a) and injection system (b). V1-V2, Valco six port gas sampling valves; M , Whitey metering valve (SS-3LRS4); T , Whitey toggle valve (SS-1GS4); PG , absolute pressure gauge.

with helium as the carrier gas. The columns are housed in a single oven that is maintained at 100°C. The detectors are operated at a temperature of 100°C, and the bridge currents are 120 mA (argon carrier) and 250 mA (helium carrier). Carrier gas flow rates through the columns are approximately 20 cc/min. A scrubber column (5-Å molecular sieve—¼-in o.d. × 8-in. long) is positioned in front of the 10-ft molecular sieve column to absorb those gases that would affect the separating characteristics of the column, namely, CO_2 and H_2O. It has been our experience that replacement of the scrubber column every month protects the 10-ft molecular sieve column from deterioration. Under the stated chromatographic conditions, the analytical column retains the required resolution for at least six months.

Samples are injected into the chromatographs using air-actuated Valco six port gas sampling valves (GSV). The samples are manually introduced into evacuated, constant temperature, constant volume injection loops at a known pressure. Figure 1 is a diagram of the introduction system, along with the valve position at injection. A quality absolute pressure gauge is required to measure the pressure in the injection system. This laboratory uses Wallace & Tiernan absolute pressure gauges, which have a range of 0–800 mm of Hg, a scale graduated in millimeters, and a quoted accuracy of 0.5 mm.

The gas chromatographs are calibrated daily using three quantitative standard gas mixtures that were certified by the National Bureau of Standards. Response factors, based on peak areas, for each gas are adjusted daily to yield the known volume percent concentration for each species in the standard. Calculations are performed using the external standard method of analysis (Young, 1975). All samples and standards are run at a pressure of 200 mm Hg (absolute) in the injection loop. Table I, columns 3, 4, and 5, lists the component concentration in each laboratory standard along with the estimated uncertainty at the 95% confidence level, as stated by the Bureau of Standards.

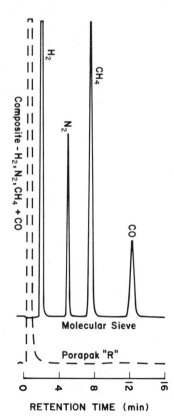

Fig. 2. Chromatogram of laboratory standard no. 1.

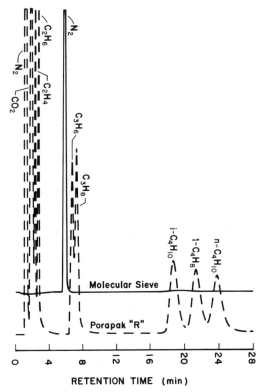

RETENTION TIME (min)

Fig. 3. Chromatogram of laboratory standard no. 2.

Near baseline resolution between components is achieved and analysis time per sample is about 25 min. Figures 2–4 are chromatographic traces obtained, respectively, when analyzing laboratory standards nos. 1, 2, and 3. These three standards have been designated as "primary" by our laboratory, and any commercially prepared gas mixture, before being accepted as a calibration gas or standard, is checked against them. This applies to mixtures used exclusively in the laboratory or those purchased for the on-line monitors.

A Hewlett Packard 3354 B Laboratory Automation System (LAS) synchronizes sample injection with the operation of the analog to digital converters, processes the data generated from the chromatographs, and generates an analytical report. A block diagram showing the interaction between LAS hardware—analog to digital converters (A/D), event control module (ECM)—and the chromatographic hardware—gas sampling valves (GSV), sample controllers (SC)—is shown in Fig. 5. Only the sample controllers were designed and built at the center; all other equipment is commercially available. A schematic of the controller along with a parts list is presented in Fig. 6.

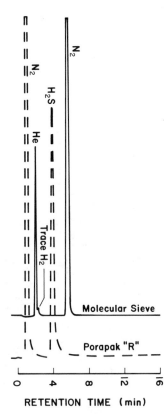

Fig. 4. Chromatogram of laboratory standard no. 3.

RETENTION TIME (min)

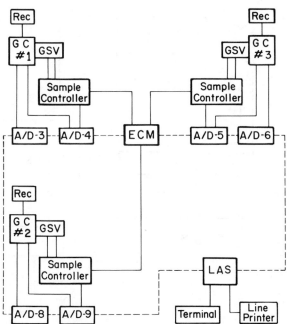

Fig. 5. Automated laboratory gas chromatographs. Rec—recorder, GC—gas chromatograph, GSV—gas sample valves, A/D—analog-to-digital converters, LAS—lab automation system, ECM—event control module.

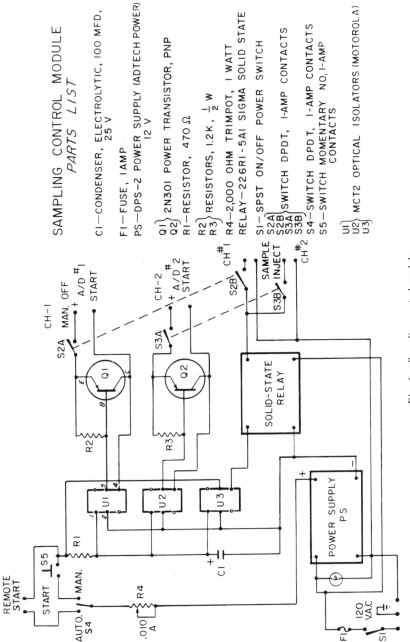

SAMPLING CONTROL MODULE
PARTS LIST

C1—CONDENSER, ELECTROLYTIC, 100 MFD, 25 V

F1—FUSE, 1 AMP

PS—DPS-2 POWER SUPPLY (ADTECH POWER) 12 V

Q1 }
Q2 } 2N301 POWER TRANSISTOR, PNP

R1—RESISTOR, 470 Ω

R2 }
R3 } RESISTORS, 1.2K, $\frac{1}{2}$ W

R4—2,000 OHM TRIMPOT, 1 WATT

RELAY—226R1-5A1 SIGMA SOLID STATE

S1— SPST ON/OFF POWER SWITCH
S2A }
S2B }
S3A } SWITCH DPDT, 1-AMP CONTACTS
S3B }

S4—SWITCH DPDT, 1-AMP CONTACTS

S5—SWITCH MOMENTARY NO, 1-AMP CONTACTS

U1 }
U2 } MCT2 OPTICAL ISOLATORS (MOTOROLA)
U3 }

Fig. 6. Sampling control module.

The analog signals from each of the chromatographs (Fig. 5) are continuously monitored by the analog-to-digital (voltage to frequency) converter modules. Each module integrates an incoming signal for 0.5 sec, digitizes it, and sends it to the computer. The LAS software sorts the information and stores it in a raw data file associated with each A/D.

In order for the LAS software to extract chromatographic data from the A/D signals, the analyst must prepare a ''method'' for each A/D module. The ''method'' is in the form of a parameter table generated by the LAS in terminal conversation with the analyst. Each A/D is assigned its own independent ''method.'' The contents of the ''method'' table include such parameters as analysis time, peak search windows, slope sensitivities, retention times, and peak names. The calculation prescribed by the ''method'' table is automatically carried out at the completion of an analysis. Peak areas, retention times, response factors, names, and concentrations are then stored in a process data file. When the calculation for both A/D's are complete, a program is called. It merges both files, sorts and then normalizes the data, and writes a report.

Through the event control module (ECM), the LAS simultaneously activates the sampling valves and starts the A/D modules.

The LAS software package allows for the automatic updating of retention times. Response factors are adjusted by the LAS software when the analyst initiates the proper routine at the completion of a standard run.

The laboratory procedure, as described, has been automated to minimize errors in the analysis of routine gas samples. The operator's responsibilities include the following: (1) introduction of the sample into the injection system, (2) initiation of the analytical run, and (3) verification of the data at the end of a standard or sample analysis. An experienced operator, utilizing the three chromatographs and faced with no major problems, can analyze about 40 to 45 samples per day.

III. Automated, On-line Gas Chromatographic System

The reader will note from the laboratory analysis procedure outlined that operator participation is minimal with respect to the actual analysis of a routine sample. This fact led the authors to investigate applications for on-line analysis. When the cost effectiveness and analytical justification of an on-line monitor can be reconciled, such a system has many advantages: (1) reduction of sampling errors, (2) elimination of sample container handling and storage, (3) minimum elapsed time from sampling to analysis, and (4) around-the-clock analysis. In the remainder of this chapter we shall discuss the steps taken to convert the laboratory chromatographic procedure to an automatic, remote, on-line analysis.

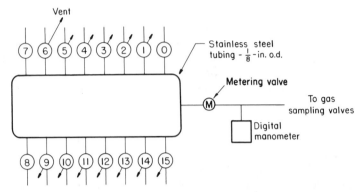

Fig. 7. Sample manifold. At position 0, 7, and 8, two-way solenoids, at other positions, three-way solenoids.

The dedicated chromatograph used in all on-line monitoring applications is the Hewlett Packard model 5730. The chromatograph, as previously described, is equipped with dual thermal conductivity detectors, gas sampling valves, carrier purification trains, and dual pen recorders.

The difference between the manual laboratory procedure and that developed for on-line monitoring lies in the sample introduction system. The laboratory procedure utilizes a "static" sampling system, whereas the on-line is designed for a "flow type" or "dynamic" sampling system.

An on-line system, in order to be used to its maximum, should have the capability of sampling more than one gas stream. Multistream selection may then require a number of standards for calibration owing to changes in components or component concentration. Therefore flexibility is desirable in the sample stream selector. Also, the sample introduction system must ensure against observable contamination between gas streams.

As the reader is probably well aware, valves for multistream selection are commercially available. Investigation by the authors revealed the following design limitations: (1) streams are selected in a step-wise inflexible order; (2) the user must devise a way of discerning the position of the valve in order to track sample selection if all inlets are not utilized; (3) all streams must vent to a common exhaust, which may or may not be advantageous.

The authors, unable to reconcile the limitations of the available stream selectors, designed and constructed a sampling manifold as shown in Fig. 7. The sample manifold, constructed of ⅛-in o.d. stainless steel tubing, describes a closed circular path with branches to each of the 16 solenoid valves and a common exit. The process gas streams of interest are connected to the manifold by three-way solenoid valves (Skinner "B" Series-B13DK1100). They are labeled as valves 1–6, and 9–15 (Fig. 7). All streams vent to a common exhaust manifold, as indicated by arrows, when the solenoids are in the deenergized position. Calibration standards and control samples are connected to the manifold by two-way solenoids (Skinner "B" Series-B2DA9250), which are closed in the deenergized position. They are labeled as valves 0, 7, and 8 (Fig. 7). The two 6-port gas sampling valves on the chromatograph are connected in series with the output of the sample manifold. A digital manometer (Ashcroft-Type No. 7580) is connected at the exit of the sample manifold. Adjustment of the metering valve (M) allows the close matching of the pressure of the sample stream to that of the pressure of the calibration standard prior to injection. The digital output of the manometer is monitored by the LAS via the ECM and, immediately prior to injection, is recorded by the LAS. A correction is automatically applied in the calculation for differences in pressure. Immediately after injecting a sample into the chromatograph, a user program selects the next stream to be analyzed; thus, the next sample stream flushes the manifold during the time required to analyze the injected sample. The design of the closed loop manifold ensures its complete flushing by the sample of interest.

The design of the sample control module (Fig. 6) allows for sample injection

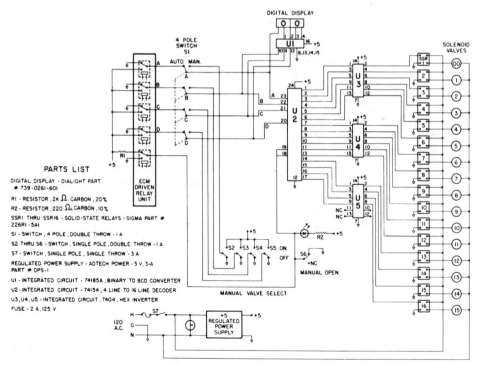

Fig. 8. Sampler stream selector.

in either the manual or automatic mode. In the automatic mode, the starting signal is initiated from the ECM; and in the manual mode, the operator initiates the signal by means of a pushbutton.

The stream selector, capable of activating any of the 16 solenoids in any sequence in either the manual or automatic mode, is presented in Fig. 8. The manual mode was designed for troubleshooting the system when difficulties arise and for use at startup to check and allow adjustments of sample or standard flows to the gas sampling valves. In the automatic mode, the stream selector is controlled by a user program. The program contains a table that associates valve number with stream name and type and designates the valves to be used in automatic operation. The analyst, through a separate user program, schedules the activation of the valves in the order desired. The analyst has the option of entering up to 128 valve designations into a schedule, and the program will execute the schedule in cyclic fashion.

The initial calibration of the on-line chromatographic system is accomplished by the analyst's analyzing the proper standard(s) and then initiating the proper

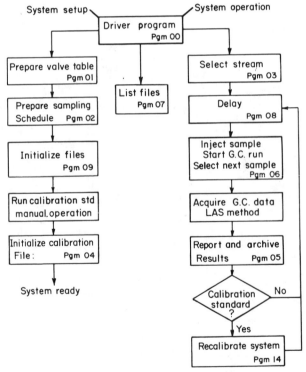

Fig. 9. Flow chart of user programs (Lab Basic II) for the automatic, online chromatographic system.

routine contained in the LAS software. During automatic operation, the updating of response factors is initiated by a user program at the completion of a standard run.

The on-line system when operating in the automatic mode has 11 user programs (Lab Basic II) associated with it. A flow chart of the programs is presented in Fig. 9. Listings of the programs are available from the authors.

IV.   Specific Applications

A.   *Monitoring Feed and Product Gases from Six Fischer–Tropsch Reactors*

1.   SYSTEM DESCRIPTION

An on-line chromatographic system was used to monitor the feed and product gas streams from six bench scale Fischer–Tropsch reactors (Baird *et al.*, 1980). A simplified schematic of the reactors involved along with the sampling points of interest is presented in Fig. 10. Sampling is done on the original feed gas, the filtered feed, and the products of reaction. The pressure of the streams at the sampling points is about 0.5 atm gauge. Each sampling point is fitted with a manually adjusted metering globe valve that allows adjustment of the sample flow to the sample manifold (Fig. 7). The feed and product gas streams have an approximate volume percent composition as shown in Table II (columns 2 and 3). An appropriate standard gas mixture (Table II) is run at selected times; response factors and retention times are automatically updated. Sample analysis

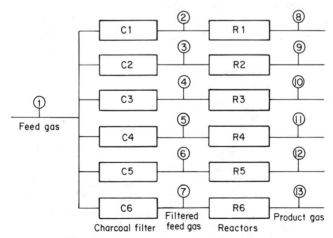

Fig. 10.   Reactor system and sampling points.

TABLE II

COMPOSITION OF GAS STEAMS—FISCHER-TROPSCH REACTORS

	Volume percent composition		
Component	Feed gas stream	Product gas stream	Standard no. 4
H_2	64.0	60.0	63.2
N_2	1.0	1.0	0.6
CH_4	1.5	10.0	8.9
CO	32.0	10.0	9.3
CO_2	1.5	15.0	14.1
C_2H_4		0.4	0.3
C_2H_6		2.0	1.9
C_3H_6		0.3	0.3
C_3H_8		0.5	0.8
$i\text{-}C_4H_{10}$		0.2	—
$1\text{-}C_4H_8$		0.2	—
$n\text{-}C_4H_{10}$		0.2	0.3
$2\text{-}C_4H_8$		0.2	0.3

time is 24 min, and all stream composition values are reported to the nearest 0.1 vol %.

The chromatograph used in this application is identical to that previously described. The chromatogram of standard no. 4, used for calibration, and the chromatographic conditions are presented in Fig. 11. The dotted line represents the elution pattern from the molecular sieve column, and the solid line represents that from the Porapak "R" column. The gases analyzed by the system and also their retention times on the particular column on which they are measured are shown in Table III. Note that there are two retention times listed for ethane. Ethane is measured on the Porapak column at 1.75 min, but it is also eluted from the molecular sieve column and creates a measurable peak at the concentration noted. However, at an analysis time of 24 min, the position of the ethane is such that it does not interfere with the current analysis on the molecular sieve column. In addition, no peaks were observed for the other hydrocarbons at the stated concentration.

The initial calibration of the on-line chromatographic system is performed by the analyst as previously described. Updating of response factors, during automatic operation, is controlled through a user program when standard no. 4 is analyzed. There are four gases (oxygen, 1-butene, iso-butane, and hydrogen sulfide) that are reported but are not present in standard no. 4. Though oxygen is not to be expected in any of the streams, it is monitored as a check of overall system integrity. Response factors for three of the four gases are set thusly:

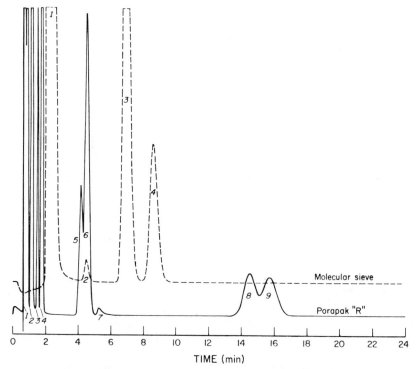

Fig. 11. Chromatogram of standard no. 4.———curve: peak 1, composite; peak 2, carbon dioxide; peak 3, ethylene; peak 4, ethane; peak 5, propylene; peak 6, propane; peak 7, trace water; peak 8, *n*-butane; peak 9, 2-butene.- - -curve: peak 1, hydrogen; peak 2, nitrogen; peak 3, methane; peak 4, carbon monoxide. Chromatographic conditions: Helium flow rates, 20 cc/min; argon flow rates, 15 cc/min; column oven temperature, 100°C; sample valve oven temperature, 100°C; helium detector, 100°C, 270 mA; argon detector, 100°C, 120 mA.

oxygen is assigned the same response factor a nitrogen; 1-butene is assigned the same response factor as 2-butene; and iso-butane is assigned the same response factor as *n*-butane. This approach is reasonable because of the close similarity in the thermal conductivities of the pairings. The response factor for hydrogen sulfide was determined by analyzing a standard blend purchased for this purpose. But the hydrogen sulfide response factor is updated by changing it according to the relative change in the propane response factor when standard no. 4 is analyzed. Thus when this standard is run, every response factor is updated.

A typical analytical report, printed at the terminal in the plant at the end of every analysis, is shown in Table IV. It contains the date and time the sample was injected, the name of the stream analyzed, the valve number of the stream, and the pressure in the sample loops at injection, namely 745.8 mm Hg (absolute). The hydrogen-to-carbon monoxide ratio and the prenormalized volume

TABLE III

RETENTION TIMES FOR FISCHER–TROPSCH PRODUCT GASES

	Times (min)		
Component	Molecular sieve column	Porapak "R" column	
H_2	2.10	< 1.0	⎫
O_2	3.04	< 1.0	⎪
N_2	4.40	< 1.0	⎪
CH_4	6.60	< 1.0	⎬ Composite
CO	8.10	< 1.0	⎭
CO_2		1.15	
C_2H_4		1.50	
C_2H_6	44.0	1.75	
H_2S		3.0	
C_3H_6		4.2	
C_3H_8		4.7	
H_2O (not reported)		5.5	
i-C_4H_{10}		10.5	
l-C_4H_8		12.5	
n-C_4H_{10}		14.4	
2-C_4H_8		16.1	

percent total (PNT) are listed. The PNT is the sum of the individual volume
percent values before normalization and is an indication of how well the system
is performing. From our laboratory experience, the PNT at the 95% confidence
level has a distribution of 100 ± 4 vol %. The report also identifies components
and lists their normalized volume percent.

TABLE IV

ANALYTICAL REPORT—FISCHER–TROPSCH PRODUCT CHECK STANDARD

4:56:12 On Oct. 30, 1979		Product check standard
H_2	63.4	
CO	9.4	Valve #7
CO_2	14.2	
CH_4	9.0	745.8
C_2H_4	0.3	
C_2H_6	1.9	H_2/CO 6.7
C_3H_6	0.3	
C_3H_8	0.8	
2-C_4H_8	0.3	PNT 99.2
n-C_4H_{10}	0.3	

2. System Performance

The performance of the system was evaluated by analyzing two commercially prepared, quantitative gas mixtures.

The procedure utilized is as follows. A cylinder of synthesis gas ($H_2/CO = 2$) was analyzed under laboratory conditions (see Section II). This cylinder was run against Laboratory Standards 1, 2, and 3 (see Table I). The results are found in column 2 of Table V. This cylinder, used as a control sample, was mounted on the automatic system together with calibration standard no. 4 (Table II). No updating of response factors was allowed to occur after initial calibration using the above standard no. 4. Over the next ten hours, a series of interchanging runs was made on the synthesis gas and standard no. 4. The analytical results are presented in columns 3 and 4 of Table V. Column 5 of Table V lists the vendor values of standard no. 4. All values in the table represent the mean of the stated number of replicates (n), and the imprecision listed is that of one standard deviation.

The excellent reproducibility and repeatability of the system under actual sampling conditions is demonstrated by the data presented in Table V. For our purposes, "reproducibility" is defined as the agreement of the analytical data produced when two or more instruments analyze the same sample. "Repeatability" is defined as the agreement of replicate analytical data generated by a

TABLE V

System Performance Chart—Fischer-Tropsch System

	Volume percent values			
	Syn. gas laboratory conditions $\bar{X}$	Syn. gas auto. syst. vs std. 1 $\bar{X}$	Std. 4 auto. syst. $\bar{X}$	Std. 4 vendor values
H_2	66.0 ± 0.11	66.3 ± 0.12	63.1 ± 0.13	63.2
N_2			0.6 ± 0.03	0.6
CH_4	1.5 ± 0.02	1.5 ± 0.03	9.0 ± 0.08	8.9
CO	31.2 ± 0.13	30.9 ± 0.14	9.4 ± 0.07	9.3
CO_2	1.3 ± 0.03	1.3 ± 0.03	14.0 ± 0.12	14.1
C_2H_4			0.3 ± 0.03	0.3
C_2H_6			1.9 ± 0.05	1.9
C_3H_6			0.3 ± 0.03	0.3
C_3H_8			0.8 ± 0.04	0.8
$n\text{-}C_4H_{10}$			0.3 ± 0.04	0.3
$2\text{-}C_4H_8$			0.3 ± 0.04	0.3
	$n = 6$	$n = 8$	$n = 16$	

sample on the same instrument. The flush time of 24 min is sufficient for purging the sample manifold, as evidenced by the absence of hydrocarbons in the synthesis gas (column 3) and by the consistent CO_2 values in the synthesis gas even though the preceding sample was an order of magnitude higher in CO_2 concentration.

B. Monitoring Gaseous Effluents at Trace Levels from Pilot Plant Coal Gasifiers

1. SYSTEM DESCRIPTION

To support the "Trace Effluent Studies from Gasifiers" project being conducted at PETC, an automatic on-line monitoring system was designed. Low level molecular nitrogen was the gas of interest in the study. Oxygen was monitored as a check of overall system and sampling intergrity. A schematic of the gasifier, along with the two sampling points of interest, is present in Fig. 12.

The gasifier was initially evacuated and then purged with helium to remove residual air. The purging of the gasifier was monitored by sampling the gas line at the helium source and then at the exit of the gasifier (Product Gas—Fig. 12). When no nitrogen was detected in the exit stream, coal, steam, and oxygen were introduced into the gasifier and the unit brought to steady-state temperature (Forney *et al.*, 1977). Molecular oxygen and nitrogen were then monitored in the product stream at selected time intervals.

Although a Hewlett Packard model 5730 chromatograph was used in the application, only one of the thermal conductivity detectors was utilized. The column used was a 10-ft × ⅛-in. o.d. stainless steel, 5-Å molecular sieve (60/80 mesh), with helium as the carrier gas. For maximum sensitivity the gas sample valve was fitted with a sample loop of 0.5 cc volume, and the analysis was run at a pressure of approximately 850 mm Hg (absolute) in the sample loop.

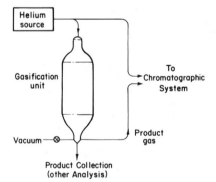

Fig. 12. Gasifier system (trace effluent studies).

TABLE VI

System Performance Trace Nitrogen Study

	Parts per million			
	Vendor values[a]		Automatic system	
Component	Std no. 5	Std no. 6	Std no. 5	Std no. 6
O_2	—	—	—	—
N_2	977	497	978 ± 5	492 ± 5
He	Bal.	Bal.	Bal.	Bal.
			$n = 12$	$n = 12$

[a] The estimated uncertainty at the 95% confidence level is ± 1% relative.

There were two sampling points of interest on the gasifier (Fig. 12) and two standard gas mixtures used in this study. The circular sample manifold (Fig. 7) was reduced in size to accommodate the four streams. The product gas stream was connected to the sample manifold by a three-way solenoid valve; the two standard gas mixtures and the helium purge gas were connected to the manifold by two-way solenoid valves.

The sample controller and stream selector operated in the manner previously described. Two commercially prepared standard gas blends (standard no. 5 and standard no. 6) were used in this application; their values are listed in Table VI. The 977-ppm N_2 standard was used to calibrate the response factor for nitrogen. The other blend was used as a control sample. Oxygen was assigned the same response factor as nitrogen. The system was set to perform an analytical run every 15 min. The duration of a test was about 3–3½ h.

2. System Performance

The performance of the system was evaluated by analyzing standards no. 5 and no. 6 under actual operating conditions. Interchanging runs were made on standards no. 5 and no. 6 over a period of six hours, and the results are presented in Table VI, columns 4 and 5. The values represent the mean of 12 replicate analyses, and the imprecision is one standard deviation. The data demonstrated both the excellent reproducibility and repeatability of the system under analytical conditions.

The authors would note that in this particular application where sampling is most critical and contamination problems so obvoius, the intergrity of the experimental data warrants the expense of an on-line monitoring system.

C. *Monitoring Gaseous Effluents from Bench Scale Coal Gasifiers*

1. SYSTEM DESCRIPTION

In order to be able to investigate gasifier startup parameters, gas species immediately generated upon the introduction of coal to a bench scale coal gasifier were studied. The description of the experiment follows. The gasifier was purged with a mixture of nitrogen and steam and then brought to operating temperature and pressure. Selected and previously prepared coal samples were then injected into the hot zone of the gasifier, and the product gases were monitored by a single stream, on-line chromatographic system. Though oxygen is not expected in the streams analyzed, it is monitored as a check of overall system integrity.

The chromatograph used in this application is identical to that previously described. The chromatogram of standard no. 7, used for calibration, and the chromatographic conditions are presented in Fig. 13. The retention times for H_2S and H_2O (vapor) on the Porapak "R" column are 0.80 and 1.15 min, respectively. Near-baseline resolution is achieved for the components of interest under the conditions described. Sample analysis time is 2 min. The volume percent compositions of standards no. 7 and no. 8, used for calibration, are presented in Table VII, columns 2 and 3.

Since there is only a single stream to analyze, there is no need for a stream selector or sample manifold. The single stream can be the product gases from the reactor or either of the standards used for calibration. Stream selection is manually accomplished, and the system and valves required to establish a flow pattern are presented in Fig. 14.

The calibration of the chromatograph must be accomplished manually by the operator directing the flow of the desired standard to the chromatograph (see Fig. 14 for valve selection). The response factor for each of the gases of interest is updated by having the operator call the proper routine in the LAS at the completion of a standard run and prior to the start of an experiment. As previously mentioned, oxygen is assigned the same response factor as nitrogen. At the completion of an experiment, the operator then reruns the standards to determine whether the response factors remained within established limits.

During gasifier operation a user program controls sample injection and also generated the analytical report.

2. SYSTEM PERFORMANCE

The performance of the single stream, on-line chromatographic system was evaluated by analyzing standard no. 7 as a sample. The excellent repeatability of

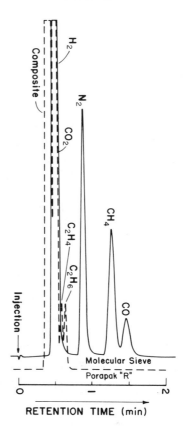

Fig. 13. Chromatogram of standard no. 7. GC Conditions: column oven temperature, 140°C; column 1—molecular sieve (5 A), 60/80 mesh, 4½ ft × ⅛-in. o.d. s.s., argon carrier gas; column 2—Porapak ''R,'' 100/120 mesh, 6 ft × ⅛-in. o.d. s.s., helium carrier gas; argon flow rate, 30 cc/min; helium flow rate, 30 cc/min.

TABLE VII

CALIBRATION STANDARDS—BENCH SCALE GASIFIER

Component	Vendor values volume percent		Automatic system volume percent
	Std. no. 7	Std. no. 8	Std. no. 7
H_2	24.9		24.9 ± .10
CH_4	10.2		10.2 ± .05
CO	9.9		9.9 ± .05
CO_2	14.9		14.9 ± .05
C_2H_4	0.51		0.51 ± .01
C_2H_6	0.50		0.50 ± .01
H_2S		0.49	
N_2	Bal.	Bal.	39.1 ± .10
			$n = 29$

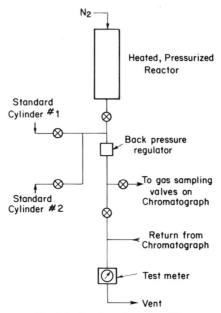

Fig. 14.   Bench scale coal gasifier.

the system is demonstrated in Table VII, column 4. The data represent the mean values of 29 replicate analysis and an imprecision of one standard deviation.

V.   Conclusions

On-line monitoring of gas streams generated from process development units in both pilot plant and experimental environments has been achieved using automated research-quality gas chromatographs. Three applications, each of which highlights one of the benefits derived from on-line analysis, have been shown. The first application maximizes chromatographic usage and allows for the unattended, around-the-clock monitoring of six research reactors. The second application minimizes contamination in trace gas analysis. The third application eliminates sample handling and minimizes the elapsed time from sampling to the generation of the analytical report. The quality of the analytical data in all three applications was as good as, if not better than, that obtainable through laboratory techniques.

The design of the sample manifold prevented contamination between streams, minimized standard or control sample waste, and allowed for a maximum of 16 different streams to be analyzed.

The design of the automated sample stream selector allowed maximum flexibility in stream selection and also gave easy access for manual monitoring during startup and troubleshooting.

When the chromatographic procedure is amendable, a single, dedicated, and automated chromatograph can increase the problem solving capabilities of the analyst through the use of the sample manifold and stream selector.

ACKNOWLEDGEMENTS

The authors wish to thank Mr. Ralph Boyce of PETC for his expertise in the design and construction of the sample controllers and stream selectors and for his concern and interest. They also wish to thank Dr. Joseph A. Feldman of Duquesne University for his valuable aid in the preparation of this report. Disclaimer: reference in this report to any specific commercial product, process, or service is to facilitate understanding and does not necessarily imply its endorsement by the United States Department of Energy.

References

Baird, M. J., Cobb, J. T., Haynes, W. P., and Schehl, R. R. (1980). *Ind. Eng. Chem., Pro. Res. Dev.* **19,** 175–191.

Forney, A. J., McMichael, J., Haynes, W. P., Strakey, J. P., Gasior, S. J., and Kornosky, R. M. (1977). Sythane gasifier effluent systems, PERC/RI-77/4. National Technical Information Center, U.S. Department of Commerce, Springfield, Virginia.

James, A. T., and Martin, A. J. P. (1952). *J. Biochem.(Tokyo)* **50,** 679.

Varian Associates (1977). "G. C. Applications Library, 1959 to 1975." Varian Associates, Palo Alto, California.

Young, I. G. (1975). *Am. Lab. (Fairfield, Conn.)* Feb., pp. 27–36; June, pp. 34–44; Aug., pp. 11–21.

5

Process Liquid Chromatography

R. A. MOWERY, JR.

Applied Automation, Inc.
Bartlesville, Oklahoma

I. Introduction

Most analysts are familiar with laboratory liquid chromatography, but few are familiar with process liquid chromatography (PLC) and its capabilities for on-stream analysis. This situation is expected to gradually improve as more process chemists and engineers become aware of the capabilities of PLC. At present, the use of PLC must be considered to be in a developing stage; however, PLC has made significant advances since the first PLC units were installed in 1973 by Applied Automation, Inc. Nevertheless, it is predicted that PLC will not come into full maturity before the mid-1980s (Frost & Sullivan Inc., 1979). This is unfortunate, since there are today industrial streams that could be analyzed by PLC (Fuller *et al.*, 1979a,b; Mowery and Roof, 1976a,b; Mowery, 1977, 1980a,b; Roof *et al.*, 1980).

The information contained within this chapter is based solely upon the laboratory and field experience of Applied Automation, Inc. However, since it appears to be at present the sole manufacturer of PLC, its experience should be representative of this developing industry.*

---

*Until recently, DuPont Instruments was manufacturing a PLC instrument. Bendix Corporation has also produced a small number of prototype analyzers for a particular customer, and the Foxboro Company has expressed an interest in PLC.

II. The Chromatographic Process

A quick review of the chromatographic process is in order since such a review provides the beginner with a few of the basic principles and terms that are common to the language of the chromatographer.

First of all, what is the technique of liquid chromatography (LC)? It is a physico–chemical method of separating the various components of a sample into fractions or bands for analysis. It is based on the phenomenon that under identical conditions each component of a mixture will interact differently with its environment. This environment consists of two phases, one a stationary phase, and the other a mobile phase or carrier. The stationary phase is a fixed bed of solid particles which may or may not be covered with a liquid coating. The mobile phase is a fluid media that transports the sample mixture past the fixed bed or stationary phase.

As the mobile phase permeates through the stationary phase, each component that was injected into the carrier distributes itself between the mobile phase and the stationary phase to a different degree. Components that are soluble or have greater attraction for the stationary phase transverse the column at a lower rate than those components that favor solubility in the mobile phase. The net result is that the various components of the injected mixture are separated into individual elution bands for analysis.

Both process and laboratory LC can be divided into several general types. Even within a type or form of chromatography there are different separation techniques that can be applied. As discussed in the following sections, some of these techniques have become so widely used that additional subcategories have evolved.

A. Liquid–Liquid Chromatography

The first important form of liquid chromatography is termed or categorized as liquid–liquid chromatography. This is partitioning chromatography in which a separation is achieved by the partitioning of the sample between the liquid carrier and a liquid stationary phase that coats the solid packing material. For stability, the liquid stationary phases that are used in PLC are chemically bonded to the packing material. This eliminates the problem of the liquid carrier washing off the stationary phase. Since one of the prime requirements for PLC involves long-term column stability, bonded phases are always used for liquid–liquid chromatographic applications.

Liquid–liquid chromatography can also be further subdivided into normal phase and reversed-phase partitioning systems. Normal phase means that the carrier is less polar than the stationary phase whereas reversed-phase systems involve a carrier that is more polar than the stationary phase. This latter partitioning system has become so popular that it is often referred to as "reversed-phase LC." In fact, three-fourths of the laboratory LC separations involve reversed-phase partitioning, and it is often the initial method for a new separation in PLC. One primary reason is that reversed-phase partitioning separates many organic compounds with water as the major carrier solvent. Economically this is not only attractive to the laboratory chromatographer but also highly desirable for any process application. An aqueous reversed-phase carrier is also generally less inflammable than most normal phase carriers. Perhaps the one slight drawback in using a reversed-phase system in PLC is the tendency for the analysis time to be longer than it is for other forms of partitioning chromatography (note Fig. 1).

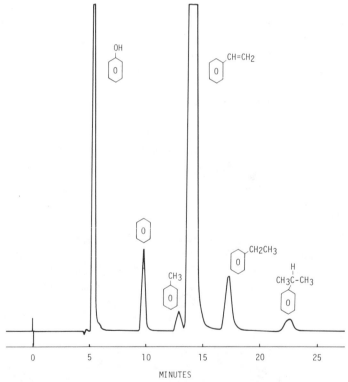

Fig. 1. Reversed-phase separation of selected hydrocarbons. Sample, as shown. Sample size, 2 μliter; column, 25 cm × 0.63 cm, S.P.—ODS, 5u (EP); temperature, 50°C; carrier, 50% CH_3CN and 50% H_2O pressure, 600 psi; flow, 0.76 cc/min; detector, UV at 254 nm; range (A), × 2.0; recorder sensitivity, 10 MV; chart speed, 5 min/in. [Reproduced from Mowery (1980a) by permission of publisher.]

A reversed-phase system has its greatest utilization in the separation of organic molecules, particularly those hydrocarbons that differ only in their carbon number. In general, paraffins require a high content organic carrier or a nonaqueous carrier for proper separation; whereas, many aromatic hydrocarbons can be separated using aqueous blended carriers with a higher percentage of water. A reversed-phase system may or may not be useful for the more polar hydrocarbons, many of which are the common petrochemicals. On the other hand, if the separation involves very polar compounds, a reversed-phase system may be ideal with an additional modifier in the carrier. This additional modifier causes either ion pairing or ion suppression of the polar sample.

In the case of ion pairing, the additional modifier is also ionic and combines with the sample to produce a less ionic paired species. Once this occurs, the retention time is more dependent upon the hydrocarbon portion of the molecule with results similar to those seen for nonpolar hydrocarbons and reversed-phase LC. To date, ion pairing has had limited application in PLC since it is difficult to maintain the necessary column stability for long periods of time.

Ion suppression is very similar in that a modifier is added to the carrier that controls the pH and causes the ionic components of interest to exist mostly in their neutral form. Once this occurs, standard reversed-phase techniques can be employed (see Figs. 15 and 16).

B. Liquid–Solid Chromatography

Liquid–solid chromatography or adsorption chromatography is the oldest form of chromatography; in fact, it predates gas chromatography by 44 years (Day, 1897; Martin and Synge, 1941). Despite this fact, efficient and timely LC separations had to wait until the late 1960s and the introduction of the modern high performance packing materials.

A separation using adsorption chromatography involves the competition of the sample components between active adsorption sites of the packing material and the solvent molecules. As in laboratory LC, silica is the most popular adsorption material for PLC (note Fig. 2).

Adsorption chromatography is most useful for organic molecules with an intermediate molecular weight. Organic compounds with a low molecular weight, assuming they are thermally stable and have sufficient volatility, are generally best separated by gas chromatography (GC). In the same way, organic compounds with a high molecular weight are generally not suitable for adsorption chromatography. Adsorption chromatography is also normally not useful for ionic compounds since many of these compounds tail appreciably. Furthermore, many ionic compounds tend to show what is termed a "nonlinear isotherm" resulting in non–Gaussian shaped peaks and retention times that are dependent

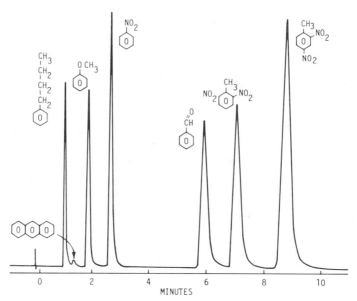

Fig. 2. A normal phase separation with a silca column. Sample, as shown. Sample size, 2 μliter (point injector); column, 12.5 mm × 0.63 mm, SI60, 5u (EP), temperature, 50°C; carrier, 5% Cl_2CH_2 and 95% hexane; pressure, 600 psi; flow, 1.9 cc/min; detector, UV at 254 nm; range (A), × 2.0; recorder sensitivity, 10 MV; chart speed, 2 min/in.

upon the concentration of the component. In a few applications, adsorption columns have been used with a water carrier to separate polar compounds; however, it appears that the separation mechanism is really a form of partitioning chromatography rather than true adsorption chromatography (Mowery, 1977).

With the exception of a carbon column, adsorption chromatography is not appreciably useful in the separation of organic homologs or compounds that differ only in their aliphatic composition. In contrast, adsorption chromatography is quite useful for separating mixtures of similar compounds. For example, aromatic isomers that differ only in the location of the functional group can be quite often separated with an adsorption column. Adsorption chromatography is also useful for separating organic compounds with different functional groups. For example, a polar compound such as nitrobenzene is retained longer than relatively nonpolar compounds such as butylbenzene (note Fig. 2).

Carbon has been used as an adsorbent for many years in GC. Its use in PLC is still at a developing stage; however, carbon has shown a unique ability to separate many hydrocarbons by molecular size, shape, and structure. Many of the branched aliphatic hydrocarbons can be separated from their isomers using a carbon absorption column. The carbon column can also sometimes be used in a reversed-phase mode with an aqueous carrier blend.

In the choice of carriers for use with carbon as an adsorption column, the stronger solvents (solvents that decrease the retention time of an eluting solute) appear to be linear molecules rather than simply solvents with a polar functional group. The interaction of carbon is through dispersion forces and is more dependent on the molecular size and shape than on the dipole moment of the molecule. Littlewood (1970) has mentioned similar interactions in the use of carbon columns in GC. n-Hexane and 1,2-dimethoxyethane are both linear molecules in which the 1,2-dimethoxyethane would be considered the stronger solvent in most LC analyses; however, with a carbon column both solvents behave as if they were strong solvents. In contrast, weak solvents (solvents that increase the retention time of an eluting solute) appear to be solvents that are highly branched or are small condensed molecules. For example, 2,3-dimethylbutane is a good highly branched weak solvent whereas n-hexane is an excellent strong modifier for adjusting retention times.

At present, spherical activated carbon (5–10 μm) does not appear to be commercially available for preparing high-performance columns for the laboratory or PLC. In PLC, the activated carbon currently must be prepared by milling irregularly shaped particles and screening the results to 5–20 μm before packing the column. The resulting columns have a relatively low porosity; fortunately, a column 5 cm in length is sufficient for many separations and does not exceed the pressure limitations of the process instrument.

A carbon column will probably not become a general purpose column for PLC, since some organic compounds are well retained and others are only removed with difficulty; however, it does provide some unique separations that are not readily obtainable with other columns and forms of chromatography.

C. Ion Exchange Chromatography*

Ion exchange chromatography, as the name implies, involves almost exclusively the separation of ionic compounds, commonly in an aqueous media. The packing material is typically a highly permeable ionic resin manufactured from the polymerization of styrene and divinyl benzene with a suitable functional group. To date, porous ion exchange resins have been found to be of limited use in PLC. This may seem somewhat surprising to the laboratory chromatographer; however, many of the ionic compounds which historically have been separated by ion exchange resins can be separated by several of the new bonded-phase ion exchange materials. This generally results in a more efficient separation, shorter analysis time, and a more stable column system for PLC.

Some of the porous ion exchange resins also show significant changes in their

*See also Chapters 1 and 3.

performance characteristics after a week or two of continuous operation. Furthermore, this is particularly noticeable when the temperature control zone or oven is operated above room temperature, which should be the case for a more efficient separation. In general, this is also true of most forms of LC in which a column that is operated between 40–60°C appears to provide the most efficient separation at a reasonable temperature.

The greatest potential for process ion exchange chromatography (PIEC) involves inorganic compounds; surprisingly enough, the demand for such analyses has been slow to develop. Nevertheless, the potential for PIEC is real (e.g., steam boilers, nuclear reactors, and power plants) and will probably involve the more stable bonded ion exchange materials.

D. Ligand Exchange Chromatography

The use of ligand exchange chromatography for process applications should also be mentioned. Ligand exchange chromatography also uses an ion exchange resin and might therefore be considered as a form of ion exchange chromatography. The basic difference is that the separation involves the exchange of a ligand which is attached to an anion or cation rather than the direct exchange of an anion or cation that characterizes ion exchange chromatography.

To date, process ligand exchange chromatography has been limited to the analysis of sugars with a straight water carrier (Mowery and Roof, 1976a,b). Even in this case, many sugars can also be separated with an aqueous-acetonitrile carrier on a stable bonded-phase column. As previously indicated, such a column is generally preferred for most process applications. For this reason, the use of ligand exchange chromatography for on-line analysis is expected to be limited to those process applications that cannot be separated by other forms of chromatography, or where the composition of the mobile phase is a consideration. This latter criterion is especially important in many of the food industries (including the sugar industry). Many companies would rather use a straight water carrier and sometimes have poorer results than risk even the remote possibility of food contamination from a mixed aqueous–organic carrier.

E. Size-Exclusion Chromatography

Size-exclusion chromatography is also known by several other names including steric exclusion chromatography, liquid exclusion chromatography, gel filtration chromatography, and gel permeation chromatography. All of these terms refer to essentially the same technique in which the sample components are separated according to their molecular size. Molecules that are smaller than the average pore size will spend a significant amount of time within the pores of the packing material whereas larger molecules will spend less time within the pores. The net result is a molecular size or in many cases a weight distribution chromatogram in which the larger molecules are eluted first, and the smaller molecules have longer elution times.

The application of size-exclusion chromatography to the process field may be the single most important form of PLC. This is primarily owing to the fact that size-exclusion chromatography is an ideal method of analyzing many polymer streams. The application of size-exclusion chromatography to the polymer field is discussed in greater detail in Section XXIV.

III.   Liquid Chromatography versus Gas Chromatography

The question, How does process liquid chromatography differ from process gas chromatography? is often asked by the process chemist or engineer who is unfamiliar with PLC.

Around 1956, the first process gas chromatographs were installed on-line to provide the process engineer with the capability of measuring stream composition (Dudenbostel and Priestly, 1956; Fuller, 1956). Process liquid chromatography expands these capabilities to include other processes which are difficult to analyze with process gas chromatography (PGC) or by other process instrumentation. One estimate is that 80% of the known organic compounds cannot be analyzed by gas chromatography (Snyder and Kirkland, 1974a) whereas many have the potential for separation and analysis by LC. In a sense, PLC should be considered as a complementary tool rather than one competing with PGC.

There is considerable overlap between the two chromatographic techniques since many compounds can be analyzed by either method. In these cases, PGC would at this time be the preferred method since it is more advanced in its development and would provide an analysis with less cost and effort. Moreover, technical personnel for instrument and system maintenance must be for the most part trained in the particular characteristics of PLC. Prior experience in PGC is most helpful but insufficient for maintaining a PLC instrument. For these reasons, most PLC applications are those where PGC or other process instrumentation cannot provide the analysis or where the use of PLC can provide a unique separation or analysis.

In many cases PLC is involved with the more complex industrial streams, some of which contain strong acids, bases, or salts. In other cases, the components of interest have high boiling points and would require an excessive temperature to vaporize. Most process gas chromatographs cannot operate continuously or satisfactorily when their oven or temperature control zone is above ca. 175°C. This is because at such temperatures the life span of most injection and

column valves is drastically reduced; furthermore, such analyzers are more difficult to service and place additional limitations on the detector system. For example, a high oven temperature reduces the sensitivity of the thermistor detector and
increases the noise level with the hot wire filament detector. Some streams also
contain volatile and inflammable components in a higher boiling-point matrix
that limits the use of "hot" analyzers in Division 1 areas (areas that contain
hazardous concentrations of inflammable gases).

Quite often, in order to comply with the requirements of a Division 1 analyzer,
the use of PGC is temperature-limited. The operating temperature, or more
correctly, the surface temperature of the heating element for the oven can not
exceed 80% of the ignition temperature for the gases present at the analyzer site
(National Fire Protection Association, 1978). In some cases PGC uses a separate
explosion-proof housing, which contains a heating element and a column,
mounted within the oven compartment. This is a sealed unit that provides sufficient heat to operate the column system at a much higher temperature while the
injection valve is maintained at a lower operating temperature. However, the use
of a multicolumn configuration within the sealed unit is limited since the switching valve has much the same temperature limitations as the injection valve.

As a rule of thumb, the upper temperature at the surface of the heating element
(at 80% of the ignition temperature) is limited to ca. 160–175°C, which trans-
lates to ca. 155–170°C for proper regulation and control of the oven compartment with many petroleum products (see Table I).

Fortunately, many hydrocarbons as well as many petrochemicals that require

TABLE I

THE IGNITION TEMPERATURE AND BOILING POINT FOR SELECTED NORMAL HYDROCARBONS[a]

Hydrocarbons[b]	Boiling point (°C)	Ignition temperature (°C)
Pentane	36	260
Hexane	69	223
Heptane	98	204
Octane	126	206
Nonane	151	205
Decane	174	210
Dodecane	216	203
Tetradecane	253	200
Hexadecane	287	202
Octadecane	317	227
Eicosane	344	232

[a] Table values as found in the National Fire Prevention Association (1977).

[b] Most common petrochemicals have a higher ignition temperature.

high temperatures for proper analysis by GC can be analyzed by LC at a relatively low temperature (typically 50°C). The use of lower temperatures generally results in the intermolecular forces between molecules being more effective in LC compared with GC. Quite often, these increased intermolecular forces can provide some unique separations with LC.

Other compounds that are of interest for PLC include compounds that are thermally labile or react on heatings. For example, peroxides such as cumene hydroperoxide (Mowery, 1977) tend to decompose, or compounds such as styrene tend to polymerize on heating. Two other areas that are difficult or impossible to analyze by GC involve the separation of ionic species and the ability to obtain a molecular weight distribution for polymers; however, both analyses are possible with LC.

PLC also employs detector systems that are different from those used in PGC, which in some cases provides PLC with increased selectivity for a given analysis.

IV. Requirements and Objectives of Process versus Laboratory Liquid Chromatography

As described by Mowery and Roof (1976a), some of the requirements and objectives of PLC are best illustrated by noting some of the differences between laboratory and process LC. The laboratory instrument is designed for manual operation in a relatively safe, stable environment, whereas the process liquid chromatograph is designed for automatic and unattended operation, long-term stability, and a dedicated analysis.

PLC objectives are to perform an on-stream repetitive analysis of the process and use the formation to control or correct the process. It must be designed for a relatively harsh environment with a wide variation in ambient temperature. It must be weather resistant, explosion proof and must meet the other requirements of Class 1, Groups C and D, Divisions 1 and 2 as outlined in the National Electrical Code (National Fire Protection Association, 1978). Furthermore, the laboratory instrument is designed for versatility, with frequent changes of operating parameters such as programming or gradient elution. In contrast, a process instrument is designed for a dedicated analysis and instrument stability. For example, the oven or temperature control zone is usually much heavier or massive than those found in the laboratory instruments. This additional thermal mass provides excellent temperature control. In fact, control within one hundredth of a degree centigrade is possible for many process chromatographs.

Another major difference between PLC and laboratory LC is in the use of columns. In PLC the emphasis is on multicolumns and valves to obtain an analysis within a useful time span (see Section XII). Time requirements are almost always a consideration in PLC, for unless the separation or analysis reflects the current status of the process, it is useless or of limited value for process control.

Gradient elution, as with other programming techniques, is a method wherein the bands' migration rates are changed during the course of the separation. This is somewhat analogous to the temperature programming techniques commonly employed by the laboratory gas chromatographer. In this case, gradient elution is the programming of the mobile phase or liquid carrier where the composition of the mobile phase is changed during the course of the separation. This is normally a continuous change or gradient.

Programming by gradient elution requires in most cases an excessive time cycle before the initial equilibrium of the column system can be reestablished. For this reason, gradient elution has not become generally accepted for PLC. It is also sometimes more difficult to reproduce a quantitative analysis with a gradient; however, some "step-wise" gradient elution techniques have been used for back-washing the column (see Section XII,D).

The process chromatographer also places more emphasis on solvent selectivity than his laboratory colleagues. This is because the laboratory chromatographer is faced with many one-time separations that do not justify the cost or effort involved in fully optimizing the selectivity of the carrier system. In most cases the proper choice of solvent combinations is difficult and usually requires an educated trial and error approach.

In PLC, as well as other routine analyses involving large numbers of samples, it is important to optimize the selectivity of the carrier since the initial cost and effort of development is repaid by a shorter analysis time. Moreover, the analysis time is often critical for process control. There are also generally fewer components of interest in an on-line analysis, and carrier selectivity does not need to be optimized for as many components as are found in a typical laboratory separation. Thus, a process separation requires the "correct" isocratic blend for a few components rather than a gradient carrier that would tend to optimize the separation of the greatest number of components at the expense of a longer cycle time. (The term "isocratic" means a mobile phase that is constant and does not change its composition during the course of the separation.)

Another difference is found in the life expectancy of the process column. A reasonable life span for a laboratory column is a thousand injections. This generally means a life expectancy as low as a few months, depending upon the usage, the type of column, and the application. In PLC, a thousand injections might be analyzed in less than a week. The point is that at least a 10-fold increase in the life expectancy of the process column is desirable before the analysis is put on-stream.

Because in PLC a longer life expectancy for its columns is desired and often required, trace components within the sample stream or carrier are more significant than those found in most laboratory operations. In general, minor amounts of impurities that accumulate on the column will almost always change the characteristics of the column and affect the analysis. Therefore, it is extremely important for these minor components to be accounted for or removed after each cycle before a stable column system can be achieved for PLC.

Even in the sales' requirements, there are some differences between laboratory LC and PLC. The laboratory instrument is sold as hardware only, but a process instrument is often sold as the hardware plus the application engineering for a specific analysis. The analysis is often considered as an integral part of the process instrument, where the problem of making the separation is transferred from the user to the process instrument manufacturer. This single fact has a considerable effect on the cost of the process instrument. For this reason, it is highly desirable for the user to perform at least a laboratory separation before considering PLC. This should be a necessary condition, although certainly not a sufficient condition, for determining whether PLC should be employed on a given process. The point is that an initial laboratory separation by the potential user demonstrates the possibility of PLC for a given analysis without the unnecessary involvement of the process instrument manufacturer. In many cases, funding, secrecy agreements, and other legal problems must be resolved before the instrument manufacturer can even determine the feasibility of the PLC analysis.

A few PLC users have begun to realize these facts and have started to employ their own analyst for process control. In many cases, these analysts can identify various streams that are applicable for PLC, perform an initial feasibility study, and even perform some optimization studies. These latter studies need to be done by an experienced chromatographer on a process instrument which is set up as a "pseudo" laboratory instrument. This type of instrument will usually pay for itself. Unfortunately, this level of expertise is the exception rather than the rule. In most installations, a PLC instrument is maintained by personnel whose level of expertise is noticeably less than that of those in the laboratory. This means that a successful application of PLC is one that includes provisions for the proper training of the instrument personnel to support the analyzer system.

There are also some differences in sample preparation between PLC and laboratory LC. In the laboratory, as many analysts know, one manually filters or cleans up the sample prior to the injection. PLC requires that these functions be automated. In many cases, the sample preparation system is more complex and expensive than the actual LC analyzer. Even the simplest stream requires some sample preparation. It may be nothing more than filtering the particulate matter; however, without a clean representative sample the analysis would quickly become useless. In practice, the instrument is only as accurate as the sample it receives.

Sample preparation systems may also include other automated hardware components, including sample dilution, when necessary. The techniques of sample dilution will be discussed in greater detail in Section VII. Sample dilution is needed where the sample concentration exceeds the linearity of the detector or where it is desirable to dilute the sample prior to the injection. For example, viscous polymers must be diluted before they are injected onto a size-exclusion column.

In the preceding paragraphs we have noted several differences between process and laboratory LC instruments. The PLC user should be aware of these differences and avoid the temptation of using a laboratory instrument where a process instrument is required. There are significant differences in design between the two types of instruments and one type should not be substituted for the other.

V. Sampling Systems

The technology of sampling systems is complex, therefore, for the purpose of this chapter, the discussion must be limited. This does not mean that sampling systems are not important; for in fact they are a vital part of process chromatography.

There are standard configurations that have been used for years in PGC, many of which have been adapted to PLC. However, PLC is still developing, and many of the new applications require new sample system designs to meet the specific needs of liquid chromatography. Therefore, it is difficult to describe a "typical" sampling system for PLC.

Despite this difficulty, all sampling systems must perform automatically certain functions. These functions include

(a)   The removal of a representative sample from the process;

(b)   Regulation of the pressure and temperature on the sample provided to the analyzer;

(c)   Prevention of any vaporization or loss of the sample;

(d)   Quantitatively, transportation of the sample to the analyzer;

(e)   Provision of a means of introducing, when necessary, a calibration sample or standard;

(f)   Provision of a means of introducing, when necessary, an internal standard, a deferred standard, or even a reactive mixture for sample derivatization;

(g)   The ability to switch between sample streams without cross-contamination in a multistream application;

(h)   Provision of a means of returning the sample to the process or to a waste container;

(i)   Removal of dirt and any other extraneous material from the sample; and

(j)   Provision of a quantitative means of sample dilution when necessary.

As can be seen, there are several functions that must be accomplished before the sample ever reaches the injection valve. Each function is important since a malfunction in any one area is likely to cause an erroneous analysis. Sample filtering and dilution are particularly critical functions and are discussed in greater detail in Sections VII and VIII.

VI. Multistream Applications

A multistream application requires a stream switching system that automatically connects one stream at a time to a single PLC. It is usually controlled from the programmer (see Section XX,D), which selects a different stream each analysis cycle. Any stream switching system should be able to switch from one stream to another without introducing pressure or flow changes. Furthermore, it must not produce any cross-contamination between streams or have large interior volumes that must be swept out before a representative sample can be provided for the analyzer.

Figure 3 shows one such system which uses a "double block and bleed switching scheme" to switch between three streams without cross-contamination. In this diagram a calibration sample line is also provided; however, it is controlled by a pair of manual blocking valves (V6,V7). V1 through V5 are typically electromagnetic valves.

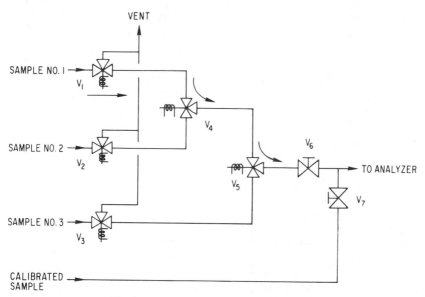

Fig. 3. Typical three-stream switching system.

In Fig. 3, the ''common port'' of each valve is the ''right-hand port'' with the ''normally opened port'' located at the top of each valve diagram. This means that the normally opened port is connected directly to the common port without any valve activation; in the same manner, the ''normally closed port'' is connected to the common port only on valve activation. When sample stream 1 is being analyzed only V1 is activated. An analysis of sample stream 2 requires the activation of valves V2 and V4 whereas the activation of V3 and V5 is required for the analysis of stream 3. In general, the total number of valves required in a ''double block and bleed'' switching system equals twice the number of streams minus one.

VII. Sample Filtering

Filters are the simplest sample preparation system and are widely used to remove small particulate matter from the sample. A PLC sample filtering system is basically the same as that used in PGC. In most process applications, a bypass type of filter is used. In a bypass filter, the filter elements are tangential to the flow of the sample stream, whereby the sample stream produces a swirling action within the filter that tends to remove the filtered particles from the filtering elements. In this way, the filters are self-cleaning, and a clean filtered sample is drawn tangentially for analysis. An occasional drawback with the bypass type of filter is its high flow rate requirement. The filter does require a sufficient sample flow rate past the filter elements in order to operate at its full potential. The bypass filter is generally located in the sample preparation cabinet just below the analyzer. Figure 4 is one example of a hardware configuration that might be found within the sample preparation cabinet.

In addition to the bypass filter, the sample is also filtered before reaching the injection valve. This latter filter is often used with a 0.2-μm filter element and is placed in front of the injection valve for maximum capability of providing a clean injection. It is also an inline type of filter that calls for periodic replacement during routine maintenance service.

VIII. Sample Dilution Systems

In addition to filtering, many samples require dilution prior to injection. Sample dilution is considered an integral part of a sample preparation system. It has been our experience that several different dilution systems are required for PLC. Figure 5 shows one such dilution system which is ideal for low dilution ratios. In most cases, these samples are nonviscous petrochemicals in which dilution is designed to bring the sample within a usable linear range of the detector.

Fig. 4. Sample preparation cabinet.

In Fig. 5, the sample is brought to the dilution valve and a sample loop. Upon activation of the dilution valve a fixed volume is injected into a suitable solvent. The actual dilution is obtained with two lengths of capillary tubing (R1 and R2), which provide two different flow paths. If the resistance of R1 is ten times higher than that of R2, then ten times more carrier will flow through R2 than through R1. This results in a ten-to-one dilution of the sample at the junction downstream

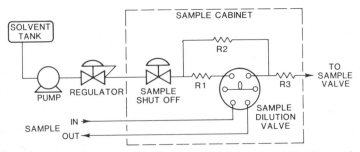

Fig. 5. 50:1 sample dilution system. [Reproduced from Mowery (1980a) by permission of editor.]

of the dilution valve. In some cases, flow controllers (see Section XI,B) are used in place of R1 and R2 in order to provide the necessary dilution.

This basic type of configuration is ideal for samples that require dilution ratios from two to one up to about fifty to one. Beyond a fifty-to-one dilution ratio, solvent consumption becomes economically unfeasible, and one is better off using a different dilution system. Such a dilution system is shown in Fig. 6.

The dilution system of Fig. 6 is designed for samples requiring higher dilution ratios. This system can economically handle a 500-to-1 dilution ratio including polymeric samples with viscosities up to ca. 20,000 cP. This type of dilution system is also preferred for samples that change viscosity during the course of production.

In Fig. 6, the dilution is obtained with a dilution chamber. The sample dilution chamber is an 800-μliter vessel that contains a magnetic stirrer. Upon activation of the dilution valve the magnetic stirrer quickly mixes the sample into a fixed volume. This arrangement provides a dilution profile with a sharp leading edge that decays exponentially after passing through a concentration maximum. Despite the exponential decaying of concentration, reproducible results within 1%

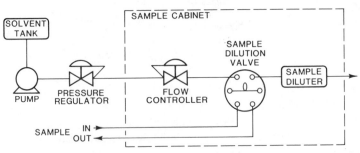

Fig. 6. 500:1 sample dilution system. [Reproduced from Mowery (1980a) by permission of editor.]

can be obtained. Best results are obtained by injecting the diluted sample near the profile's dilution maximum.

The precision obtained from the dilution system of Fig. 5 should be better since the concentration profile resembles a squared off peak. However in practice, both dilution systems can provide about the same precision. This suggests that other parameters usually dominate the overall precision of the analysis.

IX. Solvent Handling, Mixing, and Degassing

Proper solvent handling cannot be overemphasized. Most LC solvents are inflammable and must be handled with caution. An operating process LC instrument is designed for hazardous areas, whereas a filling operation that pumps or supplies the solvent into the reservoir may or may not be compatible with the surrounding environment. Therefore, local operating and safety procedures must be established by the user for this operation. In addition to safety, the carrier may also require proper blending before the reservoir is filled.

Many carriers use binary or ternary solvent mixtures that must be blended into one homogeneous solution. This latter requirement is surprisingly difficult for new users of PLC. Invariably, new users do not ensure proper blending. The results are improper retention times, resolution problems, and a drifting baseline. Once the user realizes the importance of proper blending, most if not all of these problems disappear. In some cases, it is desirable to purchase commercially available equipment for mixing the solvents. Inferior mixing techniques or inadequate solvent quality will always cause stability and analysis problems. It is also recommended that the mixed carrier be filtered just before the filling of the reservoir.

The problems associated with solvent degassing have been handled in the laboratory by a variety of methods (e.g., gas blanket, magnetic stirring of the blended carrier, heating, vacuum) whereas PLC generally employs a technique of back pressuring (e.g., 50 psig) to prevent degassing. This is commonly done with a length of capillary tubing downstream of the detector. This technique also allows the use of solvents near or above their boiling points. For example, even as little as two atmospheres of backpressure will increase the boiling point of water to about 120°C.

The use of gas blankets with an inert gas of low carrier solubility should not be ruled out. Gas blankets displace the dissolved gases within the carrier and provide a blanket of inert gas in the headspace above the carrier. This technique not only prevents dissolved gases from forming detector bubbles and interfering with the analysis, but may be necessary for the stability of certain solvents and packing materials (Leitch, 1971). Bakalyar *et al.* (1978) have shown that dissolved oxygen forms light-absorbing complexes with several solvents at 254 nm. More recently, the affect of dissolved gases at the lower wavelengths has also been reported (Walker *et al.*, 1980). In general, solvent degassing is undesirable for PLC since it requires careful control of solvent gases for weeks at a time and causes an upset condition for several hours when the carrier reservoir must be replenished.

X. Analyzer Enclosure Safety

There are several enclosure designs that meet the necessary safety requirements including the use of enclosures with flange and threaded covers. A flange cover uses a smooth surface, which no longer meets the safety requirements once the surface corrodes or has been scratched. If a flange cover is not hinged to the analyzer it may become scratched the first time the cover is removed and set on the concrete floor of the analyzer house. The flange cover is also relatively heavy and must be bolted in place, whereby entry into the enclosure generally requires effort and time to remove the necessary bolts. Furthermore, a single loose bolt can be dangerous.

The threaded cover design also has certain drawbacks. Most importantly, for a threaded cover to meet the explosion-proof safety requirements it must be engaged seven full threads (Class II fit). Unfortunately, too many maintenance technicians rarely count the number of threads they engage when replacing the cover. Instead they merely replace the cover for easy removal the next time maintenance is required. For these reasons, it has been found that the "Y-purge" enclosure is ideal as the primary safety approach for PLC. It should be noted that the threaded cover design or threaded explosion-proof housing is commonly used with the "Y-purge" enclosure as a means of isolating some of the internal system components (e.g., detector system). These internal threaded enclosures must also be engaged seven full threads before they meet the safety requirements; however, they are not necessarily opened each time maintenance is required. Nevertheless, a good maintenance practice calls for checking the internal explosion proof housings for proper thread engagement before closing the enclosure door.

The "Y-purge" enclosure simply maintains a positive air pressure within the enclosure to prevent entry of external flammable gases and to sweep out any liquid leak from within the analyzer. If, for any reason, the air pressure is lost, an alarm is activated and/or power is removed from the instrument. In the same way, as described in the following section, any significant leakage in liquid shuts off the carrier or dilution pump.

XI. Analyzer

A PLC analyzer is normally located near the process stream in a suitable structure for weather protection. The smallest buildings are termed shelters and are designed for one instrument. In most cases, a three-by-five-foot area will provide adequate shelter for one analyzer. Access is gained to the instrument by opening two large doors from either the front or the back side of the shelter. There are also various walk-in facilities that vary greatly in size, cost, and design. Some of them are true house in many respects (see Fig. 7). In all cases, they are leak-proof and form at least a semi-insulated environment. In many cases, additional insulation is also used along with heating and cooling systems to maintain a more stable environment.

Analyzer buildings are commonly constructed of twelve gauge galvanized sheet metal over a substructure of channel iron and angle iron. In some cases, in which the environment is not suited for a galvanized structure, the exterior is painted with an epoxy paint. Some structures are mounted on skids for temporary deployment at a site. More often, the buildings are permanently installed at a given site on a concrete slab.

The analyzer itself is weather resistant, explosion proof, and designed to be

Fig. 7. Inside a typical multianalyzer shelter house.

Fig. 8. Optichrom® 102 process liquid chromatograph.

located in areas containing possible hazardous concentrations of inflammable gases. It has also been designed to meet the other requirements of a Class 1, Groups C and D, Division 1 analyzer (see Figs. 8 and 9).

As can be seen in Fig. 9 this chromatograph is a wall mounted unit. It is also a second generation instrument that has incorporated many of the features and more from the earlier model of Fig. 8. Both models use, where possible, "like components" that have been adapted from twenty-four years of on-line GC experience. Perhaps, the most significant feature of these figures is that they do not look like laboratory instruments. They are not! They are process instruments and designed for a process environment.

Figure 10 shows a block diagram of a typical PLC system. Starting in the upper left hand corner one notes the carrier reservoir. A solvent reservoir is a container for storing four-to-eight-weeks supply of carrier for the PLC instrument. In most cases, these reservoirs are tanks constructed from stainless steel. A typical volume is 17 gal, which provides approximately 45 days of carrier at 1.0 mliter/min.

The reservoir is located in or near the analyzer house; however, the exact location may depend upon the temperature properties of the carrier. It may also contain pressure gauges, a sight glass, pressure relief valves or vents and other accessories for monitoring various aspects of the reservoir.

A properly blended solvent is drawn from the reservoir through a 37–74-μm filter by a carrier pump. An additional 0.2-μm filter is generally placed downstream of the carrier pump to provide additional filtering of the carrier. Next in line is a restrictor and a low pressure shut-down switch. The low-pressure shut-down switch is a safety device designed to shut off the carrier pump if there is a break in the solvent line. The function of the restrictor is to dampen pressure pulses and to reduce the volume of carrier that escapes into the surrounding environment from a line failure. This allows the low pressure shut-down switch a chance to shut off the pump with minimum loss of carrier. The low-pressure shut-down switch also turns off the pump when the solvent supply has been depleted, thus protecting the pump from wear and unnecessary high-speed cycling. Any pump shut-down must be restarted at the analyzer and can not be restarted from the programmer.

A pressure regulator follows the shut-down system. It has been found that a 2000-psi output regulator is satisfactory for most PLC applications. In many applications, the carrier pump pressure is maintained as high as 3000 psi. Thus, the pressure regulator may provide a pressure drop as high as several thousand pounds with the output pressure of the regulator typically set at around 1000 psi. This arrangement provides an extremely stable pressure source that is essentially independent of fluctuations in the carrier pump. In addition to the regulator, it should be recalled that the pump pulses are also dampened by the restrictor as well as the low-pressure shut-down switch and the pressure gauge. The net result

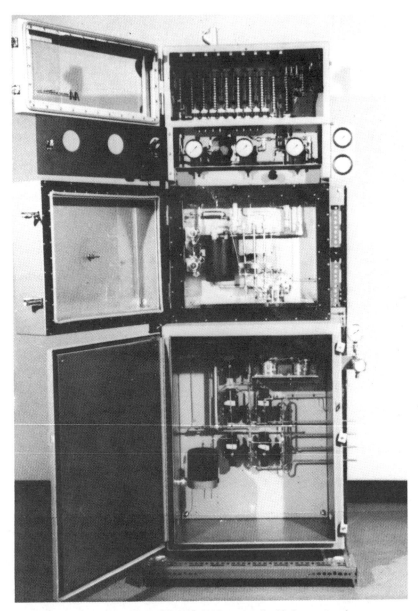

Fig. 9. Optichrom® 2100 process liquid chromatograph.

is that a stable pressure control mode is much more readily obtainable in a process instrument than with many corresponding laboratory instruments. Since solvent flexibility is not required for PLC, the process instrument can also use larger dead volume components prior to the injection valve than the laboratory LC to dampen out the pump pulses. Also, it is not as critical to use flow-through components to avoid unswept dead volumes (prior to the injection valve) in PLC as it is in laboratory LC.

In Fig. 10, the next component in the block diagram is the flow controller. In some PLC applications, flow control is not required since pressure control will provide an adequate analysis. The primary exception is size-exclusion chromatography with the analysis of polymers and other macromolecules (see Section XXIV). In polymer analysis, an error of only a few seconds in the elution time can represent an error of thousands of molecular weight units (MacLean, 1974). Thus, very precise flow control is essential for a molecular weight distributional analysis.

The process flow controller is a differential type of controller with a preset fixed resistance. This resistance determines the flow rate through the controller.

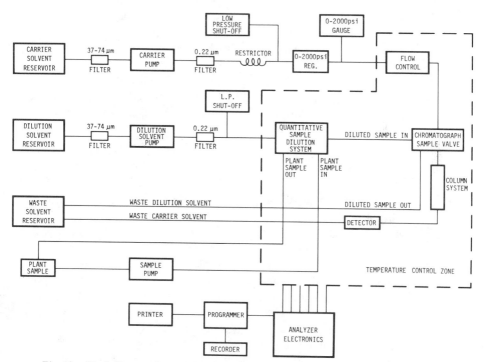

Fig. 10. Block diagram of a typical PLC system. [Reproduced from Mowery (1980a) by permission of editor.]

In practice, the pressure regulator is set to several hundred pounds or at least 30% higher than the operating pressure of the flow controller (see Section XI,B). This arrangement provides a very stable pressure source and allows the output pressure of the flow controller to vary up to the output pressure of the regulator. Normally, this range in pressure will not occur unless the system has major problems such as a "plugged" or partially "plugged" column.

From the flow controller, the carrier commonly passes through an additional filter prior to the sample valve. As indicated in Fig. 10, the sample valve injects the sample onto a column system that exits through a detector. The column system normally contains several columns with various valve arrangements. Several of these valve arrangements are discussed in Section XII.

In Fig. 10 one notes that the analyzer may also contain a dilution system. In some cases, as shown in Fig. 10, the dilution system may require a separate solvent. In may also require a separate reservoir as well as a pumping and safety shut-down system. From the pumping and safety shut-down system, the dilution solvent enters a sample dilution system such as previously described in Section VIII. In most cases the sample dilution system is located below the analyzer in a separate temperature controlled zone (sample preparation cabinet, Fig. 4). In a few cases, particularly for viscous polymers, the dilution system is located in a temperature controlled cabinet next to each reactor, whereby each polymer sample is diluted at its reactor and transported to the analyzer. These cabinet units are sometimes called "satellite sample preparation cabinets."

The remainder of the components in Fig. 10 are fairly self-explanatory: the printer, the recorder, the electronics, and programmer as well as various assorted hardware components. However, many of these components are discussed in later sections.

A. Carrier Pump

A liquid chromatographic pump provides the necessary force for the carrier flow within the chromatograph. Many different pumping systems have been checked in our laboratories and in the field. Most were inadequate for PLC, with the pneumatic amplifier pump (see Fig. 11) appearing to be the best for general service; however, a few electrical reciprocating pumps have been used.

The pneumatic amplifier pump uses compressed air at a relatively low air pressure (~ 60–80 psig) to drive a large-area piston, which in turn transfers a relatively high pressure, by amplification, onto a small-area piston in contact with the liquid carrier. As the carrier pressure approaches the maximum amplification pressure, the pump slows down and finally stalls when the carrier pressure equals the pneumatic pressure times the amplification ratio of the pump (typically 36:1). This results in the pump maintaining the desired pressure while

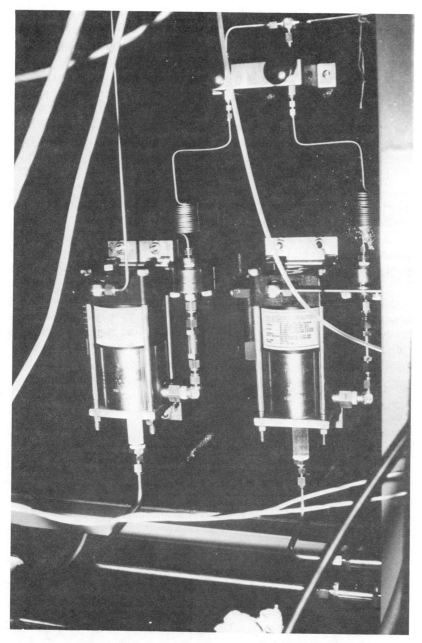

Fig. 11. Two liquid pneumatic amplifier type pumps.

consuming little if any power. At the end of the stroke, the pneumatic amplifier pump is equipped with a power return stroke that refills the pump cavity in less than 1 sec. This causes only a momentary imbalance in the liquid output pressure at the pump, which is effectively filtered out by other downstream components.

The advantages of this type of pump are that it is fairly economical to operate, with a relatively pulse-free output, and furthermore that it does not require an explosion-proof pumping motor, since air provides the necessary power to drive the pump. The pump is also self-limiting and will not exceed its preselected pressure value. By comparison, an electric reciprocating pump requires an explosion-proof motor and a pressure relief valve, which relieves the pressure with a carrier bypass loop back to the carrier reservoir.

Both reciprocating and amplifier-type pumps provide a constant flow as long as the resistance of the system remains constant. In practice, the resistance of chromatographic systems is never quite constant over an extended period of time. However, the process instrument does not rely solely on the pump for constant flow or control.

As previously noted, a pressure regulator, as well as a flow controller, may be used in PLC. The simplest control is pressure control. The use of a pneumatic pump along with the filtering of pulses by the various system components results in a very stable pressure output at the regulator. Minor pump variations due to pump cycling, check valves, "O" rings, and other pump packing materials are also effectively filtered out. For accurate flow control, an additional component, the flow controller, must be installed after the pressure regulator.

B. Flow Controller

An efficient flow controller is required for some PLC applications. One such is the Model LC221S dome-loaded flow controller (Veriflo Corporation, Richmond, CA). In operation, the output carrier from the pressure regulator is fed into the base chamber of the flow controller. From the base chamber the carrier flows through a poppet valve into a middle chamber and back out the flow controller. At this point, a link of capillary tubing connects the middle chamber to an upper chamber and provides a fixed pressure drop between the upper and middle chambers. In the upper chamber the carrier flow is simply in and out, and onto the column system or dilution system. The upper chamber also contains a spring whose tension is manually adjustable. The spring pushes against a stainless steel diaphragm that separates the upper and middle chambers.

A needle-like shaft with a flared base, which provides a variable opening between the lower and middle chambers, is also in contact with the diaphragm. This needle shaft is called the poppet valve. The tension of the spring in the upper chamber tends to hold the poppet valve open against a small fixed spring in the base chamber while the back-pressure difference across the capillary resistor tends to push back against the diaphragm and the upper spring to close the poppet valve. The net result is that the opening through the poppet valve, and thus the flow rate, is controlled by the back-pressure resistance of the capillary tubing as compared with the tension of the upper spring. Fine manual adjustment, up to approximately two-and-one-half times the flow rate, is possible by changing the tension of the upper spring with an adjustment screw. More drastic changes in the flow rate are also possible by changing the length of the capillary tubing or resistance across the flow controller.

Since the viscosity of the carrier is a function of temperature, the flow controller must be placed in a constant-temperature zone or oven. Under these conditions, and with a constant outside ambient temperature, the flow controller will control the flow rate within 0.25%. Even if a 100°F change occurred in the outside temperature, the flow controller would maintain the flow within 0.5% (assuming a constant oven temperature of ±0.3°F—see Section XI,C). In contrast, a typical pressure regulator only maintains constant flow within 3%.

Practically, the maximum allowable change in flow rate for size-exclusion chromatography is about 2%. This means that for polymer analyses the dome-loaded flow controller is ideal and can maintain the flow rate within 2%, even if there is a 30% change in column permeability. In practice, a molecular weight analyzer also uses an internal marker to correct for minor flow changes (see Section XXIV,B). Obviously, a pressure regulator should not be used by itself for size-exclusion chromatography, since the flow rate varies directly with any change in column permeability.

C. Oven

The process oven or temperature control zone has a much more massive heat sink than that normally found in laboratory instruments. Likewise, the warm-up time is also generally greater than that of laboratory units. A PLC analyzer is normally operated at a single isothermal temperature, involving a dedicated analysis. This allows the process instrument to be designed with a more stable temperature control zone. As mentioned in the preceding section, a controlled temperature is extremely important—in fact, necessary—for precise flow. It is also important for consistent column performance and can be so for detector stability.

Most of the analyzer components (e.g., columns, detector, injection valve, flow controller) are also maintained at a common isothermal temperature in a single temperature control zone. Figure 12 exemplifies an oven compartment containing several of these components.

In the figure a single column configuration is connected between the injection

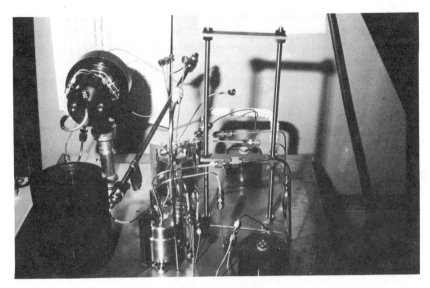

Fig. 12. Typical oven compartment for a PLC analyzer.

valve and a 254-nm ultraviolet (UV) detector (left-hand side, marked 254). There is also a column-switching valve located in the center of the oven; however, it is not in use with the column configuration shown. The two cylindrical objects on the right side of the oven are preheaters, used to bring the carrier and sample temperatures to the oven temperature prior to their entry into the injection valve. The large cap (left foreground) is a threaded explosion-proof cover for the UV detector; it must be engaged at least seven full threads to seal the UV detector housing. This isolates a possible high-voltage ignition source from possible inflammable sources within the oven (see also Section XIV and the improved UV detector model of Fig. 20). The cap also provides an additional heat sink for the detector, thus further reducing local variations in oven temperature. The oven compartment is closed with an insulated box-like cover that slips over the oven base (Fig. 12) and forms a "Y-purge" enclosure.

The oven is heated by compressed air that is passed over electrical heating coils, or in some cases cooled with compressed air to operate the temperature control zone at a subambient temperature. Since the oven is operated at a single isothermal temperature, a fixed precision resistor is used as a set point value to maintain the desired temperature. The oven is not normally operated below 60°F (16°C) or above 175°F (80°C). The major expected exception is the on-line analysis of polyethylene, which usually requires 130°C. At present, this application is in a developmental stage. In most PLC applications a 122°F (50°C) oven is ideal. With steady ambient conditions, the temperature control zone can be expected to maintain its temperature within 0.02°F (0.01°C). With ambient temperature variations from −28°F to +128°F (−33°C to +53°C), oven temperatures should still remain within 0.3°F (0.17°C) of the preset value.

Whenever possible, all electrical components for a PLC instrument should be physically separated from the oven enclosure. In most cases they are located in a separate "Y-purged" compartment. As previously noted, all electrical components that must be located within the oven enclosure are placed in explosion-proof housings. This design approach allows the oven to be readily accessible for maintenance without creating a possible ignition hazard. Furthermore, a separate electrical compartment operating at ambient temperature generally provides better results, because there is less noise, greater electronic stability, and increased operating life for the electronic components.

The PLC instrument also incorporates a temperature-limiting and shut-down feature. The temperature-limiting circuit is designed to limit the temperature of the air leaving the heater to 80% of the lowest ignition temperature for the compounds and gases present at the analyzer site. If the limiting circuit fails and the heater element reaches a certain preselected value (ca. 180°C), a safety shut-down circuit automatically shuts off the instrument and sends an alarm signal to the programmer. This feature ensures additional safety if a malfunction in the temperature-limiting circuit occurs. Likewise, a failure in the oven's compressed air supply, which heats or cools the oven, also triggers the shut-down circuit. If temperature shut-down occurs, the instrument cannot be restarted from the programmer; instead service personnel must go to the analyzer and turn off the power to the instrument. The shut-down circuit resets itself upon the reapplication of power to the instrument. However, it will not maintain power to the instrument unless the initial shut-down problem has been corrected.

D.   *Sample and Column Switching Valves*

Both the sample injection valve and any column switching valves must operate reproducibly, reliably, and automatically over many cycles. A syringe injection is not used in process chromatography. The diaphragm-plunger valve (sometimes called a "stomper" valve) used for many years in PGC, when modified for higher pressure and smaller dead volume is satisfactory in most cases for PLC. There are also several other types of valves that could be used for PLC.

The rotary valve and the slider valve are two common injection valves used in laboratory LC. Some may be adapted for PLC; however, the process user should note that their basis of operation is that of maintaining positive pressure between moving surfaces to prevent leakage. This mode of action results in wear, which limits the life of the valves; and failure of either the slider or rotary valve could cause an unsafe condition with respect to solvent leakage. Therefore, the use of

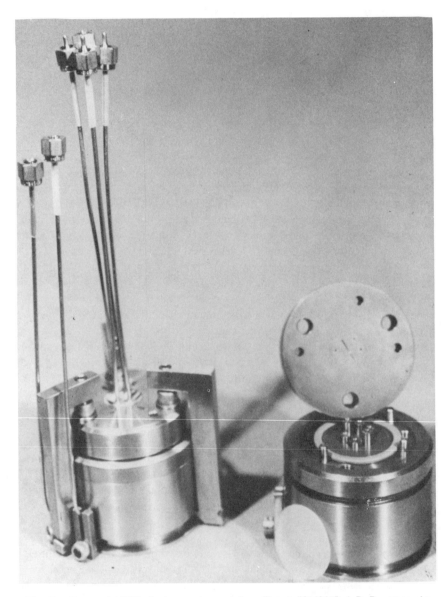

Fig. 13. Two model VIII, diaphragm-plunger valves. [Patent #3140615, A.B. Broerman, Applied Automation, Inc.; licensed by Seiscor, Division of Seismograph Service Corporation. Reproduced from Mowery (1980b) with permission of publisher.]

these valves may require that they be placed within their own explosion proof housing with a solvent sensing detector that provides an alarm signal for the control room. In contrast, the diaphragm-plunger valve rarely leaks even when the valve fails, and many a similar GC valve has operated for millions of cycles without repair. The lifetime figures on the higher pressure LC valves are less and are dependent upon the operating pressure. Broerman (1981) has reported the operation of a PLC diaphragm-plunger valve at 1000 psi for over 200,000 cycles. However, 200,000 cycles represent nearly four years of continuous service with a 10-min cycle time.

Figure 13 illustrates two diaphragm–plunger valves. The one at the left is completely assembled for use. The right hand unit has been broken down to demonstrate a few of the major components. The face of the cap is shown with a 2-μliter internal slot that forms the valve's injection volume. The cap also contains four connecting tubes that exit into the face of the cap, one of which is a capillary tube that connects the injection valve to the column system.

Figure 13 also illustrates the valve base with six plungers that have been overextended for viewing. The disklike object between the valves is the diaphragm, which is inserted between the valve base and cap. A cushion disk is also installed between the valve base and diaphragm. Its use increases the operating life of the diaphragm.

In operation compressed air (60–80 psig) is used to switch the valve, whereby the air pressure is directed between two possible chamberlike pistons. Each piston controls the movement of three plungers. Each activated piston amplifies the applied force onto its appropriate plungers using similar principles, as described for the pneumatic amplifier pump (Section XI,A). In turn, each activated plunger pushes against the cushion disk and the fluorocarbon diaphragm that seals the flow paths between alternate valve ports. This causes the various flows within the valve to follow paths of least resistance and flow between ports whose plungers are relaxed. Likewise, in valve switching, the latter plungers are activated while the previously raised plungers are relaxed. (Note the assembled valve of Fig. 13 and the two tubes in the valve base that provide the necessary air pressure paths for switching the valve.)

Compared with other valves, the diaphragm-plunger valve is also ideal for column switching in that the valve does not add appreciable "dead volume." A switching valve uses a six port cap, instead of the four port injection cap of Fig. 13, with all capillary tubing. In Section XII column switching and multicolumn configurations are discussed in more detail. Here it is sufficient to point out that such techniques require low volume switching valves as well as low volume fittings and connecting tubing. In the case of the diaphragm-plunger valve, the internal operating volume is less than 1 μliter. Of course, as with all valves, connecting capillary tubing is required.

The process user should be aware that process liquid chromatographs are designed for column switching and use low "dead volume" fittings. Therefore, any maintainance that substitutes fittings or tubing can be disastrous to the analysis if the maintainance engineer does not use similar internal diameter fittings or tubing.

The diaphragm-plunger valve is also characterized by a short activation time. Most air-operated slider or rotary valves must move about a quarter of an inch once they overcome the force of friction. On the other hand, the diaphragm-plunger valve has few moving parts that move about ten thousandths of an inch. This means that the plungers in Fig. 13 are indeed over extended for viewing. In operation, the diaphragm-plunger valve is virtually friction-free with a total activation time of less than 150 msec.

One major difference between process and laboratory valves for LC is the need for high pressure. Typically, a laboratory valve may be rated as high as 5000–7000 psi whereas the PLC valve in Fig. 15 is rated to approximately 1500 psi. This rating could be extended by increasing the valve's activation time and operating with a higher switching pressure. However, as discussed in the next section, it is generally not necessary for PLC. This is also the case for the more recently developed Model 20 valve (Applied Automation, Inc.), a new generation and improved diaphragm-plunger valve (Broerman, 1981). The performance of this latter valve is comparable if not better than the Model VIII valve, yet it costs less and is easier to service. It was also designed for general service below 1500 psi.

XII. Liquid Chromatography Column Configurations and Considerations

Most PLC separations are run below 1200 psi. This is because PLC uses a combination of columns and column-switching techniques that must provide an economical analysis within a fixed analysis time. Many times these column combinations are shorter than a typical laboratory column. For example, many of the simple reversed-phase separations that are done with a typical 25-cm laboratory column can be done with a short 4-cm column simply by increasing the water content of the carrier and choosing a stationary phase that is "wettable" with the higher concentration of water. Many octadecylsilane (ODS) packing materials fulfill this requirement and are wettable even with 100% water. The use of a high percent of water in the carrier is also desirable in that the safety precautions and the operational cost associated with the carrier are less.

For similar reasons, the typical laboratory approach of increasing column length to improve resolution has limitations in PLC. Increasing the column length generally means an increase in the analysis time, unless the instrument can be operated at a higher flow rate with the corresponding higher instrument pressure. Conditions must be chosen in PLC that not only provide the required separation but take into consideration the limitations of the instrument and the demands of the process.

First of all, the major instrument limitations include the pressure rating of the instrument, the detector limitations, and the "economics of the carrier." The upper pressure limit is that pressure where the injection valve and the carrier valves will not function properly. Many of the characteristics and limitations of PLC detectors are discussed in several later sections. "Economics of the carrier" simply means that a flow rate must be chosen that is economically feasible to operate the instrument with a given carrier yet provides the necessary analysis time. In practice, an economical flow rate is usually less than 2 mliter/min. In most cases carrier flows of between 1 and 1.5 mliter/min are most useful and are commonly used in PLC.

As previously indicated, a major constraint is the maximum allowable analysis time, which is often dictated by the requirements of the process. The analysis time must be within predetermined limits to be useful for controlling or monitoring the process. In general, although the maximum number of plates available for a separation can be limited by the instrument and the economics of the carrier, it is almost always limited by the maximum allowable analysis time. The maximum allowable analysis time for a process separation, other than size-exclusion separations, is almost always less than 15 min and is probably less than 10 min.

The result of these limits is that the separation must be more nearly optimized with selective solvents and column-switching techniques than is found to be the case in laboratory LC. Often, short columns are used that provide better solvent economics, less column dilution, shorter analysis time, and lower pressure requirements. In the same way, spherical packing materials with an above ambient oven are commonly employed to reduce the operating pressure of the instrument. Furthermore, for monitoring or controlling most process streams, generally one to three components are monitored. Owing to these factors (unlike laboratory LC), most of the components in a sample stream are not monitored or separated but are merely back-flushed or back-washed off the column.

The following sections describe several standard column configurations that are widely used in PLC, most of which have been adapted from PGC. These examples have been restricted to configurations involving a sample injection valve (SV) and a single column switching valve (CV). However, column configurations involving as many as four valves have been used in PLC. This appears to be a reasonable upper limit, since the addition of valves always increases the void volume, which eventaully becomes significant and affects the separation. The control of void volumes are much more important in PLC compared with PGC. In most cases, the more complex valve arrangements are composites based on two simple valve configurations. For example, a heart-cut system (Section

XII,E) might be combined with the advantages of forward column stepping (Section XII,A).

A. Forward Column Stepping

Figure 14 illustrates a valve arrangement that is commonly called "forward column stepping." There are several uses for this arrangement. One is to increase the effective column length without requiring excessive high pressure to maintain an acceptable flow rate. This type of configuration is used where a slight increase in resolution is desirable. In operation two comparable length columns are used, when, after the sample components have been eluted onto column 2, the column valve is switched causing the components to pass back onto column 1. After these components are on column 1, the column valve might be deactivated causing them again to pass back through column 2 before reaching the detector. As can be seen, this is a closed loop system where the components can be cycled around through a continuous length of columns. This technique has been called "recycling chromatography" by some laboratory chromatographers (Porath and Bennich, 1962). Unfortunately, after three or four cycles, dilution and band spreading generally begin to degrade the gain in resolution.

More often, forward column switching is used in the analysis of components that have widely different elution characteristics. In practice, column 1 is usually a short column while column 2 is typically 15–25 cm. Upon injection the sample

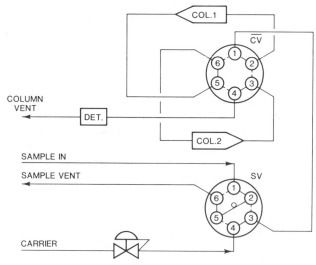

Fig. 14. Forward column stepping.

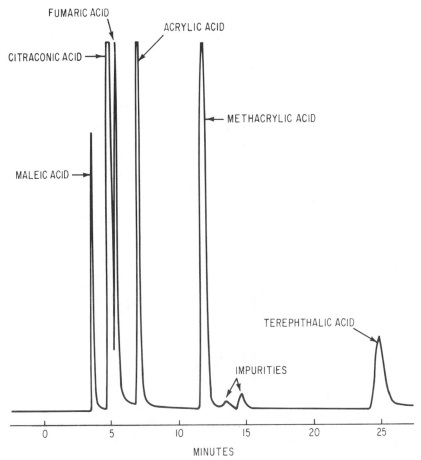

Fig. 15. Separation of several organic acids without column switching. Sample, as shown. Sample size, 2 μliter; column, 22 cm × 0.43-cm i.d. + 25 cm × 0.43-cm i.d., S.P.—ODS, 5u (EP); temperature, 64°C; carrier, 1% acetic acid, water pH 3.0; pressure, 1200 psi; flow, 1.2 cc/min; detector, UV at 254 nm; range (A), × 0.05; recorder sensitivity, 10 MV; chart speed, 5 min/in. [Reproduced from Mowery (1980a) by permission of editor.]

is placed on column 1, from which the fast eluting compounds are eluted onto column 2. At this point, column 1 is switched behind column 2 with the result that the well-retained compounds can be eluted within a reasonable time off the shorter column. This provides a possible method of separating compounds that have widely different elution characteristics without resorting to gradient techniques and unnecessarily long analysis times.

An example of the use of forward column switching is shown in Figs. 15 and 16 in which it is desirable to increase the resolution of some components,

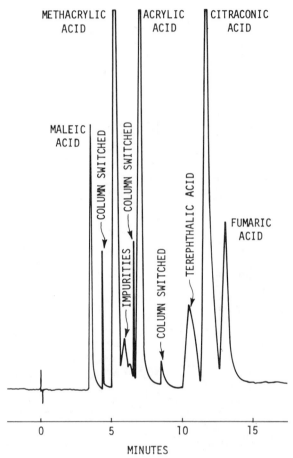

Fig. 16. Separation of several organic acids with use of forward column stepping. Sample, as shown. Sample size, 2 µliter; column, 22 cm × 0.43-cm i.d. + 25 cm × 0.43-cm i.d., S.P.—ODS, 5u (EP); temperature, 64°C; carrier, 1% acetic acid, pH 3.0, water; pressure, 1200 psi; flow, 1.2 cc/min; detector, UV at 254 nm; range (A), × 0.05; recorder sensitivity, 10 MV; chart speed, 5 min/in. [Reproduced from Mowery (1980a) by permission of editor.]

rearrange other components, shorten the analysis time, and use an isocratic carrier. Figure 15 is a typical laboratory type of separation performed with two columns and the forward column stepping configuration of Fig. 14. This application shows the separation of several organic acids, some of which are commonly used as the starting material for certain polymers. In the chromatogram of Fig. 15, the column-switching valve was not activated during the analysis. The injected sample is simply eluted through both columns before passing into the detector.

In contrast, Fig. 16 illustrates the results when the column switching valve is activated and deactivated at appropriate times. Comparing the two figures, we note that the first valve activation does not occur until the maleic acid has been eluted through both columns. Upon valve activation the following citraconic and furmaric acids are recycled back onto column 1. At this time the moderately retained methacrylic acid and tailing impurities are beginning to be eluted off column 1. They are passed into the detector without the need of column 2. The following valve deactivation causes the citraconic and fumaric acids again to be recycled back onto column 2. At the same time, column 2 is connected once more to the detector with the acrylic acid just beginning to be eluted off column 2 (note that it has the same retention time as found in Fig. 15). Finally, after the acrylic acid has passed into the detector, the column valve is again activated in order that the well-retained terephthalic acid can be eluted off column 1. The citraconic and fumaric acids are also eluted off column 1 after passing twice through column 2 and three times through column 1.

In summary, Figs. 15 and 16 show that forward column stepping can be used to increase the resolution of some components (e.g., citraconic and fumaric acids), rearrange other components, and in some cases, provide a shorter analysis time (note the analysis time difference between Figs. 15 and 16).

B. Reverse Column Stepping

Figure 17 is similar to Fig. 14 except that column 1 is reversed upon column valve activation. This arrangement illustrates a standard configuration for PLC

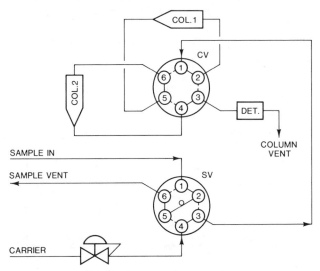

Fig. 17. Reverse column stepping.

called reverse column stepping. Reverse column stepping provides another method of separating many industrial streams that contain components that are not eluted in an acceptable time. Quite often, these are similar components that can be grouped as a single composite peak for analysis. As suggested in Fig. 17, they are simply back-flushed off the head of column 1 and into the detector. In other applications monitoring the concentration of the back-flush components is not required; however, some type of back-flushing is required to provide column stability. One such configuration is described in the next section.

C.   Column Back-Flushing

Back-flushing is another technique common to process chromatography. In its simplest form, a column is reversed causing the carrier to flow in the opposite direction. This technique allows the elution of normally well retained components in a much shorter time span. Furthermore, this technique is regularly used in PLC to remove the minor impurities that could affect column stability over an extended period of time. Figure 18 shows a configuration for a direct back-flush to vent.

Figure 18 or similar configurations probably represent the most common valve-column arrangement found in PLC. In use, column 1 is essentially a "guard column" that performs a partial separation, where, once the components of interest have passed onto column 2 (the analytical column), column 1 is back-flushed to vent. One should note that (as compared with reverse column stepping) this is a parallel operation in which column 1 or the guard column is back-flushed while most of the separation is being performed on column 2. In this

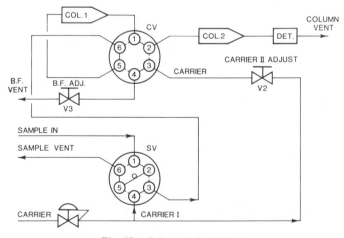

Fig. 18.   Column back-flushing.

manner the analysis, cleaning, and reestablishment of the injection equilibrium are all accomplished within a reasonable time span. For the more difficult industrial streams, in which back-flushing does not provide a stable analysis, column back-washing can sometimes be employed.

D. Column Back-Washing

Most column configurations are applicable to either PLC or PGC; however, "back-washing" is a column configuration peculiar to PLC. This technique is used for well-retained components that even back-flushing will not remove. The procedure involves the use of a precolumn or a guard column onto which the sample is injected. The precolumn allows passage of the components of interest onto a second analytical column, which makes the bulk of the separation. Once the sample has passed onto the analytical column, the precolumn is reversed or back-flushed. At this point, a cleaning solvent becomes the back-flush carrier.

Back-washing uses the same basic configuration as Fig. 18 except that an additional injection valve (called the "washing valve") is placed after the carrier split and before the sample injection valve. The washing valve is essentially a second sample injection valve except that a washing solvent is injected in place of the sample. The washing valve is normally configured with an external loop of 5 to 10 mliter. Thus, when the washing valve is activated, the carrier pushes the washing solvent through the system and backwashes the guard column to vent. This technique (a temporary step gradient) also provides an accurate washing volume since the primary carrier is used to force the cleaning solvent through column 1. The net result is that the cleaning and equilibrium of column 1 can be reestablished in a reasonable time while in parallel most of the separation is being made on column 2 with the original carrier.

E. Heart-Cutting

The technique of "heart-cutting" involves venting most of the undesirable components after a partial separation and passing only components of interest onto a second analytical column. It is commonly used in process chromatography when the analysis requires the monitoring of a minor component whose elution time follows closely behind a component of much larger concentration. The technique has the result that the requirements of the analytical column are fewer, since the heart or bulk of the major interfering component has been removed or cut. Thus, the term "heart-cutting."

In Fig. 19 most of the interfering components are allowed to flow through column 1 to vent. Just prior to the elution of the minor component, column 1 is connected to column 2. This allows the minor component along with lower

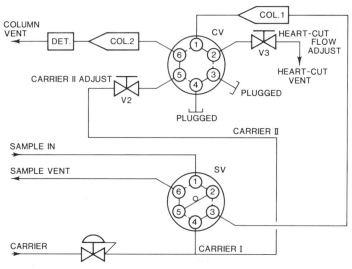

Fig. 19.   Heart-cutting.

concentrations of the interfering component to be placed on column 2. Once the minor component of interest has passed onto column 2, column 1 is again connected to vent in order to continue the venting of the major component. The net result is that since the concentration imbalance between the major and minor components is more favorable, the theoretical plate requirement for the necessary resolution is less and within the capability of column 2. Sometimes this technique also allows the length of column 2 to be reduced, with a corresponding reduction in the analysis time and the pressure requirements.

XIII.   Introduction to Process Liquid Chromatographic Detectors

In PLC, the role of the detector is to provide continuous monitoring of the concentration of the solute in a mobile phase by means of some type of output signal. The major problem with LC detection is that the physical properties of the solute are similar to those of the mobile phase. Therefore, detection of a solute has been limited to the following approaches (Snyder and Kirkland, 1974b):

1. The mobile phase must be removed before detection.
2. A property must be chosen in which the mobile phase does not interfere.
3. A bulk physical property must be employed with a suitable reference.

To date, PLC has only used detector systems based on the latter two approaches. The most common example of approach no. 2 is the UV detector that can only be used when the mobile phase and the sample do not absorb at similar wavelengths. An example of approach no. 3 is the differential refractometer which monitors a difference between the refractive index of the solute in the mobile phase as compared with the refractive index of a reference carrier.

Liquid chromatographic detectors can also be classified as either destructive or nondestructive. A destructive detector consumes the sample so that it cannot be collected for further analyses. The transport flame ionization detector is a destructive detector used in laboratory LC. In PLC, only nondestructive detectors have thus far been employed.

Another general way to classify detectors is to designate them "selective" if they respond to a limited class of compounds and "universal" if they respond to most compounds. Selective detectors such as the UV detector are generally more sensitive and less likely to be affected by minor changes in the carrier composition. Unfortunately, there are many compounds for which there is no selective detector. Both the refractive index (RI) detector and the dielectric constant (DC) detector are considered to be universal detectors since they respond to many different compounds. Sometimes they are also referred to as bulk detectors, since they monitor a change in a bulk property of the solute compared with that of the carrier (approach no. 3).

XIV. Ultraviolet Optical Absorption Detector

The most common PLC detector is the process ultraviolet detector (see Fig. 20). Its characteristics have been previously described (Walker *et al.*, 1980). In most cases the UV detector contains a low pressure mercury lamp that emits its primary radiation at 254 nm.

The reference and sample flows are through individual fluid cells designed to focus the lamp's radiation through the cells and onto a dual photodiode. In this way the absorption characteristics of the sample stream are compared with those of the reference stream. If the sample components from the column have absorption characteristics that are different from those of the reference cell, an imbalance in the bridge circuit results that is proportional to the concentration of the components in the sample cell.

At present, scanning wavelength detectors are not used in PLC. This is partly due to the relatively limited lifespan of most UV sources. For example, the common deuterium lamp is rated at approximately 500 hours. Moreover, under process conditions where the deuterium lamp would normally be placed in an above ambient environment, one would expect an even shorter lifespan. Further-

Fig. 20.   A process UV detector as mounted within the oven compartment. [Reproduced from Mowery (1980b) by permission of publisher.]

more, continuous sources have less available light energy at the particular wave-length, a higher noise level, and show poorer long-term stability characteristics than is desirable. This is not to say that different wavelengths are not useful for process applications. In fact, selectable discrete wavelength UV detectors are desirable and are used in PLC.

A discrete wavelength UV detector uses high intensity line sources, which provide a much higher illumination than a continuous source at specific wave-lengths. This results in a better signal-to-noise ratio than is possible with a continuous source, as well as the required lifetime characteristics.

The use of the UV detector with discrete wavelengths can also provide an additional flexibility, since the UV detector can become a specific detector for two closely resolved components with different optical absorption characteris-tics. At present, discrete UV wavelengths from 214 to 350 nm are available for use in the PLC optical absorption detector. Higher wavelengths into the visible region are also available. However, applications that require the use of optical detectors in the 300–700-nm region have been rare in PLC.

Despite the availability of various wavelengths, process UV detectors will probably use either the 254-nm mercury lamp, the 214-nm zinc lamp, or the 229-nm cadmium lamp as the primary UV source. As previously indicated, a wave-length of 254 nm is employed for most process UV applications. For trace analysis, a 214-nm zinc lamp generally provides greater sensitivity if the separa-tion can be made with one or more of the following solvents: hexane or a similar aliphatic hydrocarbon, methanol or similar alcohols, acetonitrile, or water. The experienced chromatographer will note that the 214-nm UV detector is readily usable with reversed-phase LC; however, it becomes more difficult to find sol-vents suitable for separating compounds by most other forms of LC. In some cases, the 229-nm cadmium lamp may be the best compromise between solute sensitivity and solvent selectivity (Walker *et al.*, 1980).

A typical optical path length for the process UV detector is 1 cm with an 8 or 23 μliter cell volume. Shorter path lengths are possible and can be employed in applications that would require some limited dilution (1:1 or 1:2) to operate the detector within its linear range.

In general, the UV detector is the preferred PLC detector if the sample has a chromophore with adequate absorbance. For a process UV detector, the absor-bance A is the same measurement unit as commonly used with the laboratory photometer, namely as expressed by the Bouger–Beer's law and the following equation:

$$A = abc \tag{1}$$

In Eq. (1), a is the molar absorptivity (liter/mol cm) at the specific wavelength, b is the optical path length in centimeters, and c is the sample concentration (mol/liter).

As can be seen from Eq. (1), the absorbance A is a function of the product ac with a fixed path length cell; therefore, the detection limits for the UV detector can be approximately related to the molar absorptivity. Generally, these detection limits range from ca. 10^{-9} g for a compound with molar absorptivity of 10^4 to ca. 10^{-6} g for a compound with a molar absorptivity of 10. In most cases, additional information is required for an exact comparison; however, many chromophores give detection limits of ca. 10^{-8} g with the UV detector.

In practice, the process chromatographer is often more interested in the minimum full scale value that can be set for a given component than in its detectability at twice the signal-to-noise ratio. The minimum desirable full scale setting for a process UV detector is 0.01 absorption units (AU) since a flowing carrier at ca. 50°C normally provides a baseline noise of around 1×10^{-4} AU. Thus, Eq. (1) can also be used to solve the approximate concentration needed to produce a 0.01 AU full scale response. For example, a compound with a molar absorptivity constant of 200 and a molecular weight of 100 would require an approximate concentration of 5×10^{-5} mol/liter, or would be present to about 5 ppm in order to produce a full scale reading of 0.01 AU at the detector. In the previous sentence, note the words "at the detector." This means that the injection concentration owing to column dilution must be larger. Column dilution is a function of the injection volume, column efficiency, and the retention volume; however, an approximate dilution factor can be calculated (Bristow, 1976) and used. In the previous example, a column system with a 20:1 dilution factor would require approximately 100 ppm to be present in the initial injection in order to produce a full scale response of 0.01 AU. Similar considerations also apply in the use of all other detectors for PLC.

XV. Refractive Index Detector

The refractive index (RI) detector is the second most popular PLC detector. There are two basic RI designs, the Fresnel or reflection type and the deflection type. Both detect the presence of the solute by the change of a refractive index in the effluent. The Fresnel reflection type is currently not being used for PLC.

The deflection type that is being used for PLC employs a single beam of light that passes twice through the detection and reference cell in series. The second light pass occurs after being reflected from a mirror where the light is retransmitted back through the cells for focusing on a pair of photoresistors. In turn, these photoresistors are part of a bridge circuit. Figure 21 shows a diagram of the deflection-type process RI detector.

All of the components in Fig. 21 are contained within an explosion-proof housing, which provides both safety and a heat sink for the detector. The light emitting diode (LED) light source, mask, and dual photo cell are located at one end of the explosion-proof housing while the fluid cells are located at the other

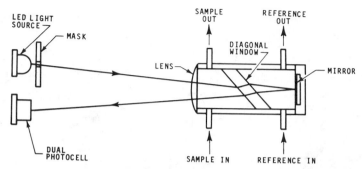

Fig. 21. Diagram of a PLC refractive index detector.

end, approximately 18 cm away. Inside and between the two ends of the housing, a 1-in thick glass plate (not shown in Fig. 21) isolates the electric components from the fluid components. In this manner, the safety requirements are met at the same time that the necessary light paths between the electrical and fluid components are maintained.

The major disadvantage of the RI detector is its extreme sensitivity to temperature changes. To minimize this problem the process RI detector is maintained in a temperature control zone whose temperature is controlled to within 0.02°F (0.01°C). Furthermore, the mass of the explosion-proof housing around the detector acts as an additional heat sink and ensures even greater temperature stability. This is important since temperature must be controlled within 1×10^{-3}°C in order to observe RI changes of 1×10^{-5} RI units. The effluent cells are also arranged so as to tend to cancel minor RI changes from flow and local temperature variations.

The best lowest detectable limit for an RI detector is approximately 1 ppm at the detector. In practice, this means that moderately retained compounds must be in tenths or milligram quantities before they can be successfully detected with an RI detector.

In general, there are some limitations to using an RI detector. Some are mentioned in the next section in the description of the dielectric constant detector. Despite these limitations, the RI detector is at present the most commonly used PLC detector that meets the criterion of being nearly universal (i.e., responds to most compounds).

XVI. Dielectric Constant Detector

The dielectric constant or capacitance detector is a recent development for PLC (Benningfield, 1979; Benningfield and Mowery, 1981). In fact, the Model 410 (Applied Automation, Inc.) dielectric constant (DC) detector appears to be the only commercially available DC detector built for LC in the United States. It

Fig. 22. The dielectric constant detector as mounted in the oven compartment. [Reproduced from Mowery (1980b) by permission of publisher.]

is marketed as a laboratory unit as well as a process LC component. Figure 22 is a picture of the dielectric constant detector mounted in the oven compartment of a PLC.

As described elsewhere in more detail (Benningfield and Mowery, 1981), the dielectric constant detector is a differential-type detector and contains both a reference and sensing cell. Each cell consists of two concentric cylinders with a clearance of 0.009 cm between the two cylinders. These two cylinders are electrically isolated and form a cylindrical flow path between the two cylinders or electrodes, with a total cell volume of approximately 23 μliter. The outer cylinder is at ground potential. In operation, the carrier flows through the wall of the outer cylinder, around the smaller inner cylinder, and back out the opposite wall of the outer cylinder.

There are several techniques for measuring the dielectric constant of a solute, including the use of a Shering or a Wein bridge (Scott, 1977a). However, monitoring the difference in frequency between the two cells appears to be a more satisfactory method for PLC. As indicated in Fig. 23, each cell forms part of a parallel inductance capacitance resonant circuit where the frequency of each cell oscillator is determined by the capacitance of the cell. In turn, the output frequency of the two oscillators is fed to a mixer which senses the difference between the frequencies of the reference and sense oscillators. In operation, the capacitance of the reference and sense cells is "balanced" with a frequency offset difference of about 1 kHz between the two cells for PLC. In the laboratory model, the offset value is larger (typically 5–10 kHz) in order to provide better solvent flexibility. This use of a frequency offset is a must since similar frequency outputs from both the reference and sense cells tend to couple and produce a detector with low sensitivity.

The DC detector has a linear range that extends about four orders of magnitude above the minimum detectable concentration (MDC). The DC detector also tends to provide about the same detectability for sample components with similar dielectric constants. For example, a nonaqueous reversed-phase (NARP) separation (Parris, 1978) involving a complex hydrocarbon stream of uncertain composition will normally furnish more meaningful results with less effort using the DC detector. Also, a size-exclusion separation of a polymer is often best monitored with the DC detector. This is because the sensitivity of the DC detector

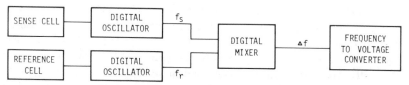

Fig. 23. Block diagram of the dielectric constant detector. [Reproduced from Benningfield and Mowery (1981) by permission of Preston Publications, Inc.]

with an injection of a polymer or hydrocarbon sample is nearly constant and individual correction factors are normally not required for quantitative results. In contrast, if an RI detector were used with the above size-exclusion separation, the apparent molecular weight distribution might be displaced from its actual distribution. This is because the response of the RI detector is not only a function of the polymer concentration but also a function of the detector's sensitivity to individual constituents of the polymer. Notably, the response of the RI detector is especially dependent upon the molecular weight once the molecular weight falls below ca. 10^4 (Barrall et al., 1968; Sadao, 1978). The DC detector also provides increased solvent selection capabilities, since many of the solvent combinations are not necessarily the same as those employed with other detectors (Benningfield and Mowery, 1981).

In many cases, the detection limits of the process DC detector are similar to or better than the process RI detector. The DC detector also complements the RI detector in applications that can not be achieved without a "universal" detector. In general, the sensitivity of a given compound with an RI detector can be related directly to the absolute refractive index difference between the compound and the carrier, whereas the sensitivity of the same compound with the DC detector is more complex.

The best detectability is attained with high dielectric constant solutes in carriers with low dielectric constants such as are common in normal phase LC. For example, acetone's detectability was found to be 0.5 ppm in a hexane carrier. Furthermore, solutes with dielectric constants greater than 8 and a linear relationship between the concentration and the dielectric constant have a predicted minimum detectable concentration (MDC) value of less than 1 ppm in a nonpolar carrier. In this case, the detectability is often better than with an RI detector.

In cases where the dielectric constant of the carrier is greater than the eluting solute, the MDC values are larger and increase as the dielectric constant of the carrier is increased. For hydrocarbons and hydrocarbon-based polymers, an MDC value of less than 5 ppm is possible with carrier solvents that have dielectric constants between 6 and 12. This range of solvents includes many of the common carriers (e.g., tetrahydrofuran and o-dichlorobenzene) used in size-exclusion chromatography. In general, using carrier solvents that have a higher dielectric constant than the eluting solute results in MDC values that are comparable to those obtained with the RI detector and a refractive index difference of 0.02. For extremely high dielectric constant carriers, especially aqueous solvent systems, the detectability is typically similar or poorer using the DC detector, as compared with the RI detector. The RI detector is also at present slightly more sensitive than the DC detector for solvent–solute combinations in which the relative differences in refractive index and the dielectric constant are approximately equal. In most cases, this involves nonpolar solvent and nonpolar solute combinations.

The principal limitations of the DC detector are basically the same as those of the bulk property RI detector. The DC detector is sensitive to changes in temperature, pressure, and carrier flow rate. Bulk detectors are also very sensitive to changes in the mobile phase composition. In addition, the DC detector cannot be used with buffers, salts, or other electrolytic solutions.

In conclusion, the DC detector has become a standard process LC detector that complements the RI detector by providing an additional "universal" detector for the process industry. Yet each detector has significant characteristic differences that allow either one or the other detector to provide better analysis for a specific industrial stream.

XVII. Electrical Conductivity Detector

The process electrical conductivity detector, like its laboratory counterpart, is primarily concerned with the detection of ionic species. In most cases, the analysis involves acids, bases, or salts and normally uses an aqueous carrier with an ion-exchange column. An electrical conductivity cell essentially obeys Ohm's law, which states that a voltage causes a current flow that is proportional to the electrical resistance between the electrodes. In a process conductivity detector, an AC potential is used rather than DC voltage since the latter causes polarization at the electrodes and confusing results.

The electrical conductivity detector is one of the simplest detectors available for process application. The detector consists of two Tefzel® cylindrical bodies which sandwich a cylindrical electrode. In operation, a short piece of stainless steel capillary tubing connects the column to one of the Tefzel bodies where the capillary tubing is joined to the Tefzel body by a platinum-lined reducing union. Tefzel is a nonconductive elastomer that provides the flow paths that, in turn, provide the electrical paths between the inlet or outlet and the center electrode.

The center electrode is a platinum-plated stainless steel disk that allows the carrier to flow through the center of the disk. Downstream of the center electrode the carrier flows out of the conductivity cell by means of another Tefzel body. This Tefzel body is identical to the inlet Tefzel body. Thus, the response of the detector is equivalent to two parallel resistors, since the detector measures both the conductivity between the center electrode and the inlet union as well as the conductivity between the center electrode and the outlet union. This design allows both the inlet and the outlet union to be held at ground potential while an AC potential is applied to the center electrode. A typical value for the center electrode is 1.6 V peak-to-peak at a frequency of 4 kHz.

This parallel resistance design offers an advantage over a two-electrode design cell in that long lengths of insulating tubing are not required to connect the detector cell to the column system. This must be a consideration in using a two-electrode conductivity cell since the impedence of the mobile phase may not be adequate to prevent ground loops that affect the stability of the amplifier and recorder. Furthermore, the addition of connecting tubing always creates undesirable dead volume that reduces the efficiency of the column.

There are several other differences between the process conductivity detector and its laboratory counterpart. For example, the process conductivity detector can withstand significantly more back pressure than many of the laboratory models. It is not uncommon for some laboratory models to leak when the back pressure exceeds about 20–30 psi, whereas a process unit can withstand back pressures of several hundred psi at above ambient operating temperatures. Furthermore, most laboratory models use simple bridge circuits to measure the difference in conductivity. The process conductivity detector employs more sophisticated electronics and uses what is termed ''phase sensitive electronics'' to make the same measurements. Simply speaking, the oscillator voltage and frequency that is supplied to the center electrode are also supplied to a field effect transistor (FET). The FET acts as a ''gating filter'' and allows only the oscillator frequency from the conductivity cell to pass. In this manner, much of the system noise is rejected, producing a detector system with better noise characteristics.

XVIII.   Other Detectors for Process Liquid Chromatography

There are several other types of detectors that conceivably could be used for PLC. Most of these have been described in the literature for various laboratory uses. These include detectors based on streaming potential, polarography, light scattering, argon ionization, scintillation, ultrasonics, electrochemical reaction, thermal conductivity, flame photometry, ion selective detectors, electron capture, and even detectors based on flame ionization (Munk, 1971; Scott, 1977b). In fact, there are many types of laboratory detectors, of which most, for one reason or the other, will never be adapted for PLC. On the other hand, both the density detector and a detector based on infrared absorption have had some limited development for possible use in PLC. At present, the only other detector worth considering for PLC appears to be the fluorometric detector. Such a detector might be desirable for some specific applications; however, to the best of the author's knowledge it has not been employed to date in PLC.

XIX.   Introduction to Programmers

A process LC, like its GC counterpart, operates automatically whereby (unless the LC is computer controlled) a separate electronic programmer is provided for each analyzer. The programmer is usually physically independent of the analyzer and located in the control room or another Division 2 area (an area where inflammable vapors are confined within closed containers or systems). Also, the readout or data presentation is also normally located in the control room, which may be several thousand feet from the actual site of the analyzer.

The programmer controls the analysis time, sample injection, selects and measures the components of interest, and converts the results for data presentation or closed loop control. The programmer must also control all column switching functions including valves used in any washing process, back-flushing, or sample dilution. In addition, scaling or calibration of each component of interest is also accomplished by the programmer or computer system. As discussed in the following sections there are three basic types of programmers that have been used for PLC.

XX.　Conventional Electronic Programmer

The conventional electronic programmer employed in PLC is essentially the same type of unit that has been used for years in PGC (see Fig. 24). Therefore, a high degree of reliability might be expected and that is indeed the case. In its simplest form, the programmer is a timer that activates certain functions at preselected times. Furthermore, as in all programmers, these functions are cyclic and must repeat automatically during each analysis cycle.

Fig. 24.　Applied Automation, Inc., Optichrom® 102, conventional electronic programmer.

The electronic programmer also contains several alarms, such as the purge alarm, which indicates a loss of purge pressure in the electronic enclosure of the analyzer. Most electronic programmers also have a power alarm, which indicates that the analyzer has lost all electrical power. In most cases these alarms are both audible and visual alarms on the programmer.

The programming of an electronic programmer may be accomplished by means of diode pins in a matrix board, where the position of the pins determines the activation time of each function. In some recent designs these functions are programmed by "thumbwheel switches." Most electronic programmers also have up to four programmable relay functions, one of which is commonly used to operate a back-flush system. The electronic programmer can also be interfaced with most computerized systems; however, as indicated in the following sections, many of the functions available to the electronic programmer are surprisingly sophisticated. It is still the most widely used type of programmer for PLC, even though microprocessor-based programmers have made tremendous advances.

A. Gating

During an analysis cycle the programmer activates a "gate" or the necessary circuitry to provide a predetermined "measurement window" in which a peak of interest is monitored for either a peak height or area measurement. This is a programmed function in which windows are turned on or off at preselected times for measuring only the components of interest. Each pair of "gates" or "measurement window" has its own attenuator that has been previously calibrated with a preset scaling factor based on a calibrated standard. In this way the data obtained for the components of interest can be presented as a direct readout in some appropriate units (e.g., mole percent or weight percent) without any subsequent calculations. Typically, an electronic programmer has up to five attenuators (ten gates) available for monitoring the components of interest.

In some applications a specialized electronics card is included within the electronic programmer where activation of the "gate" at a preselected time only activates the card circuitry. This activated circuitry must in turn sense a change in baseline slope of a predetermined magnitude before any measurement data is collected. This card is called a "slope gating card" and allows the measurement window to vary over a certain time span. In this way, the data collection is dependent upon a change in baseline slope and the elution of a peak rather than a fixed time window. It also provides the electronic programmer with some of the features normally associated with a more expensive computerized system.

B. Peak Ratioing

Peak ratios are sometimes necessary for process control. A specialty card may also be included as part of the electronic programmer for storing and ratioing two of three possible input signals; furthermore, two such "ratio/dual storage" cards can be used with the electronic programmer for PLC.

In operation, two chromatographic signals that the attenuator network has scaled are first stored and then used to calculate the ratio value. The second input signal provides the denominator for the ratio where its input value can come from either "ratio/dual storage" board.

C. Auto-Zero

The electronic programmer, like the computerized programmers, has the capability of rezeroing the baseline at preselected times. This function is called the "auto-zero." The auto-zero compensates for baseline drift by a feedback signal which adds or subtracts an offset value to the baseline. The auto-zero amplifier maintains this bias value until the next time the auto-zero function is activated. In most cases, the auto-zero function is activated each time the "gate" of a measurement window is activated for a given attenuator. The auto-zero function can also be programmed to occur at a preselected time which is independent of the attenuators.

The auto-zero requires only 300 msec to check the baseline offset and will typically correct over a range of approximately 100 mV with the electronic programmer. Some computerized programmers will correct for baseline drift up to 10 V.

D. Stream Switching

Most of the electronic programmers also have a stream switching capability in which typically four different streams can be connected or disconnected from the analyzer using only the basic conventional electronic programmer. Additional stream switching capability is also readily available by interfacing the electronic programmer with a "multistream switching unit." The latter unit can typically provide switching capabilities of up to 16 different streams for a given analyzer and analysis. Any additional stream switching capability is generally undesirable, since the time between the monitoring of individual streams becomes unacceptable for most process applications with a single analyzer. The multistream switching unit also has a number of indicator lights that show the user which stream is being analyzed and which stream is flowing through the sample valve or dilution valve.

A microprocessor-based programmer can also provide a 16-stream capability without the additional stream switching unit. Furthermore, since a microprocessor-based programmer normally controls multiple analyzers, the actual maximum capability is 16 streams per analyzer (see Section XXII). In either case, a stream identification signal will be provided to the recorder or data presentation unit. This normally precedes each chromatogram or data readout for the stream.

Both the computer-based programmers and the conventional electronic programmer with a multistream switching unit are designed so that any given stream can be easily connected or disconnected from the analysis cycle. In this way the process analyzer samples only those streams that are actually operating at any given time; in the same way, a standard reference stream can be easily connected to the analyzer for data verification and calibration.

An attenuator programming feature is also included within the computer-based programmers as well as the basic multistream switching unit. In this way, scaling with the proper attenuator can be independently programmed for each stream. This may be necessary where there are large concentration differences between the various streams. In some cases, as described in Section VIII, proper dilution techniques may also be required for some individual streams.

XXI.   Minicomputer-Based Programmers

A minicomputer-based programmer is normally only economically justified where there are a large number of chromatographs. One estimate is that a minicomputer-based system for 30 chromatographs would cost about half the price of

Fig. 25.   Applied Automation, Inc., Optichrom® 2C, minicomputer-based programmer with 6 gas chromatographs.

a similar system with conventional electronic programmers. Figure 25 illustrates one such minicomputer system.

A minicomputer-based programmer is designed to control and monitor up to approximately 60 different chromatographs involving as many as 100 different compounds. To date less than 50 commercially available systems have been installed in which only one controls a PLC. The remaining computer-based systems control and monitor PGC analyzers. However, there are several PLC instruments that are being controlled by the user's host computer system along with the general operation of the plant.

There has always been some reluctance to install minicomputer-based programmers even in PGC. This appears to be due to the high initial cost and the fear of losing the data from all chromatographs if the computer goes down. Furthermore, in recent years, microprocessor-based programmers have become available with sophisticated data reduction routines and capabilities that approach the performance of the larger minicomputer-based programmers. The next section discusses a few of the characteristics of a microprocessor-based programmer and how its characteristics differ from those of other types of programmers.

XXII. Microprocessor-Based Programmer

All major process chromatography companies offer microprocessor-based programmers for PGC; however, only the microprocessor unit as described by McCoy (1978) has been used to date in PLC. Therefore, much of the following discussion will consider this programmer as the "typical" microprocessor-based programmer for PLC (see Fig. 26).

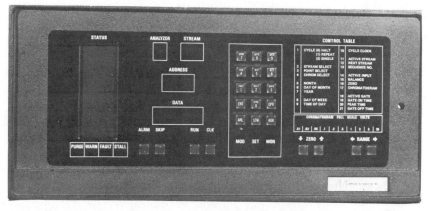

Fig. 26. Applied Automation, Inc., Optichrom® 2100, microprocessor-based programmer.

In general, the differences between the minicomputer and the microprocessor-based programmer are a matter of degree. A minicomputer-based programmer is designed to look after more analyzers and more components than a micro-processor system. A microprocessor-based programmer is in some respects a next generation minicomputer that was designed to provide sophisticated data reduction at a significant savings in cost. It has also been designed to look after fewer analyzers; however, with better "chips" being produced every year, it might be anticipated that microprocessor-based programmers will continue to expand their capabilities and functions.

A microprocessor-based programmer today typically controls a maximum of four analyzers with the possibility of 64 streams, and in some cases up to 50 different components could be monitored by a single microprocessor-based programmer. For size-exclusion chromatography, which requires extensive detector monitoring and the necessary data reduction for calculation of the molecular weight distribution (MWD), only two analyzers are at present being used per microprocessor-based programmer. A conventional electronic programmer is not used for size-exclusion separations that require an MWD.

The front panel of a microprocessor-based programmer contains a number of digital displays along with a built-in push button keyboard that allows the user instantaneous communication with the programmer. In an instant the user can call up the status of any analyzer and any stream as well as monitor any status or alarm condition.

Two types of memory are used in a typical microprocessor-based programmer. They are programmable read-only memory (PROM) and read/write or random address memory (RAM). The PROM is nonvolatile memory used to store the programs that comprise the operating system of the microprocessor. It also acts as an interpreter for the instructions that are programmed into the RAM by the user. The RAM contains the application programs for a particular system and can be altered by the user. Since the RAM is volatile, a standby rechargeable battery pack is contained on the memory board for use in the event of power failure. There is also an interface capability through which an inexpensive audio cassette tape recorder can reload the programmer's RAM within a few minutes.

Each memory board can contain up to 12 kbytes of PROM and 8 kbytes of RAM. The maximum number of memory boards is typically four, with a typical system not containing more than 16 kbytes of PROM. The typical access time for the PROM is 120 ns as compared with 300 ns for the RAM.

A microprocessor-based programmer contains all functions previously noted for the electronic programmer as well as others. Many of the functions used by the electronic programmer are preinstalled in the factory for a specific application and may require factory modification for new applications. On the other hand, both the minicomputer and the microprocessor-based programmers have many functions and options built into the software. For example, as shown in

Table II, the method of calculation for any given peak can be changed by entering a new code number into the microprocessor unit.

Also, there are other parameter codes within the microprocessor unit that can be easily changed or monitored by the user. One of these is "relay timing," which controls the activation times of all valves and functions within the analyzer. This is a time "ON" and "OFF" function and each analyzer has a maximum of eight relays that can be monitored or changed by the keyboard.

The microprocessor-based programmer also stores factors for each peak; furthermore, unlike the electronic programmer, the microprocessor-based programmer also has the facilities to automatically calculate its own calibration factors and eliminate human errors. This is done with reference samples in which the calibration factors are automatically calculated from the known concentrations of standard samples and stored for use in the analysis. In many cases the whole operation is automatic, where a standard sample is permanently connected to the analyzer. In this way the calibration factors can be automatically updated on a daily, weekly, or even a monthly basis. This can be done because the microprocessor system also contains a real-time clock that allows events to be scheduled in real time rather than just cycle time.

An additional important feature of a microprocessor-based programmer is the incorporation of a system of diagnostics. These are all alarm functions from which typically 254 different alarm codes are available. Many of these are built-in as part of the nonvolatile PROM. These codes immediately tell the operator the nature of most malfunctions. Furthermore, the purge and stall alarms have special significance with their own visual alarm lights. A purge alarm indicates that the analyzer electronic section's air purge has failed, and no longer meets the requirements of a Division 1 analyzer. A stall alarm indicates a microprocessor malfunction. Other major malfunctions normally produce a "fault" alarm and

TABLE II

METHOD OF CALCULATION OPTIONS FOR THE MICROPROCESSOR-BASED PROGRAMMER

Data code	Method of Calculation
0	Integrated area relative to zero
1	Integrated area relative to frontgate height
2	Integrated area relative to backgate height
3	Integrated area relative to the average of front and backgate height
4	Peak height relative to zero
5	Peak height relative to frontgate height
6	Peak height relative to backgate height
7	Peak height relative to the average of front and backgate height
8	Detector offset voltage, if any, of frontgate height
9	Detector offset voltage, if any, of backgate height

cause the updating of the results to be discontinued. Most alarms are warning alarms that do not interrupt the operation of the system but simply alert the operator that something needs attention. In many cases, they show that the operator has entered incorrect parameters for operating a function.

Other diagnostic and alarm features can be used by the operator to monitor the health or well being of the analyzer system. Such items as communication lines, power lines, detector balance, oven temperature, and pump pressure, can be monitored by the microprocessor-based programmer.

For logging of results, a simple printer is adequate and relatively inexpensive. Likewise, a bargraph output or trend recorder can be used. More extensive logging of information can be provided by a larger printer or even a cathode ray tube (CRT). The output information can also be furnished to a larger host computer for additional capabilities. In some cases the host computer also controls the industrial process where the chromatograph is just one of many sources that provides data for controlling and optimizing the process.

One major difference between the microprocessor-based programmer and the other type of programmers is the manner in which signals and data are transmitted. Both the electronic programmer and most of the older minicomputer-based programmers transmit analog signals over multiwire cables (approximately 22 wires per analyzer) to and from each analyzer. The microprocessor-based programmer uses an analog-to-digital converter, as well as a digital-to-analog converter within each analyzer, and transmits only digital information over serial communication lines (2 wires per analyzer). Therefore, the use of the microprocessor-based programmer can provide definite savings in cost if the communication wiring must be run over a significant distance.

Some of the characteristics of a serial communication line include a typical line speed of 1200 bits/sec with the detector signal being transmitted every 50 msec. Another characteristic is that the level of current flowing in the serial communication lines is only 20 mA.

XXIII.   Data Presentation Units

A strip chart recorder is the most common method of data presentation on which the output signal from the detector is generally recorded as a bargraph of either peak height or integrated area. Figure 27 shows a typical bargraph for a two-component sample. (Note also the excellent repeatability in the bargraph presentation with the two component sample.) Unlike most laboratory recorders, the chart drive of a process recorder is off during much of the analysis cycle. This procedure provides many chromatograms (e.g., scaled peak heights or scale heights of the integrated areas) without reams of chart paper. It is easily accomplished through the programmer and a time delay relay which activates the chart

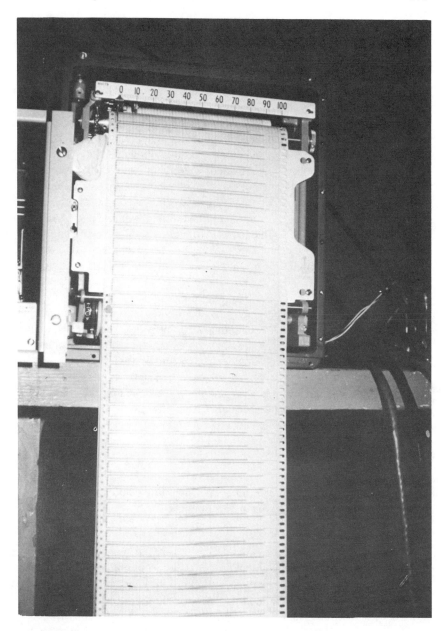

Fig. 27.   Strip chart recorder with bargraph output.

drive of the recorder at appropriate times to space the bargraph. A typical spacing of the bargraph, where the chart drive is activated, is about 8 sec for a chart speed of 1 in/min.

Trend recording is accomplished by means of a "peak picking device" or memory card, which is normally part of the programmer. The memory card is designed to measure the maximum peak height (e.g., scaled peak height or scaled height of the integrated area) for the component of interest, store this value during the analysis cycle, and present the value as an output signal to the recorder. In turn, the stored value is erased on the following analysis cycle just before the measurement gate is reactivated. In this way, the recorder pen traces a stepwise continuous line that is representative of the component's concentration.

In addition to a bargraph or trend presentation, the programmer can also provide a standard laboratory type of chromatogram, which may be required during the initial startup or routine maintenance checks. A polymer analysis also sometimes requires a visual inspection of the full chromatogram along with the various calculations from the data printout. These various presentation modes are controlled by either a function switch on the electronic programmer or through the data keyboard on a computer-based system.

Digital displays or data printers can also be used as part of the data presentation system, in conjunction with either the electronic programmer or a computer-based system. A data printer can provide a hard copy for the permanent records from which such information as component concentration at a particular time and date can be retrieved for a particular stream. The printer can also have alarm contacts in which the print out is in red if the concentration level exceeds a predetermined value. Sometimes, the data presentation unit includes a teletype or a CRT, both of which are commonly used as an integral part of a complete computerized system.

XXIV.   Process Size-Exclusion Chromatography and the Exclusion Process

As previously mentioned (Section II,E), size-exclusion chromatography is a technique in which the sample components are eluted according to their molecular size. Unlike other forms of chromatography, the stationary phase should be inert so that the retention time of the solute depends solely on the size, or more correctly, the hydrodynamic volume of the molecules. The hydrodynamic volume can be defined as that volume which is encompassed by the largest sphere that can be generated from the rotation of the molecule around its various axes.

In practice, both solute–substrate and solute–solvent interactions that affect the retention time do occur. These effects are particularly noticeable for the lower molecular weight species, where solvation or molecular association by solvent molecules can cause large changes in the hydrodynamic volume. Even changes in the environmental energy surrounding the molecules can sometimes cause additional rotational modes to occur, which in turn may affect the hydrodynamic volume.

In many cases, the solute–substrate interactions involve small polar molecules or polymers that contain very polar moieties. The correlation of elution volume with the molecular weight is best with samples whose components are a homologous series or have similar composition. For that reason process size-exclusion chromatography is usually involved with polymers of similar composition, where the retention volume can be easily related to the molecular weight, or with the separation and analysis of a small polar molecule in a sample of macromolecules.

The retention time or elution volume of a macromolecule can be related to its distribution coefficient K in which the molecule is distributed between the liquid volume contained within the pores of the substrate particles (intraparticle volume V_1) and the volume of the liquid surrounding the particles (interparticle volume V_0).

In size-exclusion chromatography, the maximum concentration of a given component within the pores of the substrate can vary from $K = 0$ to the same concentration, $K = 1$, found in the mobile phase (interparticle volume). In other words, the concentration within the pores will never exceed the concentration of the component within the interparticle volume as long as the separation involves only a size-exclusion process. In the same way, large molecules that are totally excluded from the pores of the substrate will be eluted at the interparticle or void volume V_0. In this case, the concentration of the macrospecies within the pores and the distribution coefficient are both zero. As the size of the macrospecies is gradually reduced, a point is reached where penetration of the pores by the solute occurs. This point is commonly referred to as the "exclusion limit" for the column. Any additional reduction in the hydrodynamic volume causes a column separation to occur until a point is reached at which a further reduction in the hydrodynamic volume will not increase the amount of penetration by the molecules into the pores of the substrate. This point is called the "permeation limit" of the column, where the size of the molecule allows complete and free access into the pores of the substrate ($K = 1$).

It is hoped that it has become clear that solute fractionation only occurs between the exclusion and permeation limits of the column with retention volumes ranging from V_0 to ($V_0 + V_1$); thus, for maximum separation in the shortest time, columns with large pore volumes (V_1) and lower void volumes (V_0) are best. Furthermore, since V_1 is proportional to the amount of packing material contained within the column, size-exclusion chromatography must use longer columns of larger diameter than is common with the other forms of PLC to provide the necessary resolution. These dimensions generally result in a longer analysis time than is found in other forms of LC, since V_0 is also larger.

A typical column configuration generally involves two to four columns with appropriate exclusion and permeation limits. These columns are commonly constructed with 6.2-mm i.d. tubing, 25 cm in length and packed with 7–10-μm particles. They are normally operated with column pressures of 300 to 600 psi and precise flow rates (within ¼%) of ca. 1 mliter/min. Furthermore, most process size-exclusion columns are the same commercially available prepacked columns as are used by many laboratory chromatographers. They are used owing to the reluctance of most vendors to sell bulk packing materials.

A. Polymer Parameters

Polymers are often the result of a few specific chemical reactions that are repeated many times. These reactions are generally between relatively simple organic compounds in which each reactive compound or monomer behaves as an individual unit for a "building block process" that ultimately results in the formation of a specific polymer. In turn, most specific polymers are mixtures involving a range of different molecular weight species in which many of the physical properties that characterize the polymer depend directly upon the degree of distribution within the mixture. Two batches of a given polymer may even have the same average molecular weight but different physical characteristics if there is a noticeable difference in the distribution of molecular weights. Therefore, it is important for the polymer manufacturer to be able to characterize each polymer batch for uniformity and the desired physical features. Process size-exclusion chromatography provides a method for characterizing the polymerization process on-line through a molecular weight distribution (MWD) analysis.

The following equation describes the most commonly used parameters for characterizing a given polymer:

$$\bar{M}_b = \Sigma\ N_i\ M_i^b / \Sigma\ N_i\ M_i^{(b-1)} \tag{2}$$

where N_i is the number of species with a molecular weight of M_i. When $b = 1$, Eq. (2) yields the number-average molecular weight ($\bar{M}_n$). The number-average molecular weight can be related to the colligative properties of the polymer (e.g., vapor pressure and osmotic pressure). When $b = 2$, Eq. (2) provides a value that can be related to the light scattering behavior of the polymer in solution. This latter value is called the weight-average molecular weight ($\bar{M}_w$) and can be related to such properties as mechanical strength.

The ratio of the two averages $\bar{M}_w / \bar{M}_n$ is also of interest, since it describes the breadth of the molecular distribution. This ratio is called the polydispersity of the polymer. A polydispersity of nearly 1 indicates a polymer with a narrow distribution of molecular weights.

There are other measurements that can be obtained from the molecular weight distribution (MWD) (e.g., Z-average molecular weight and viscosity average molecular weight); however, such additional measurements are normally not required to monitor the production of a given polymer with predetermined characteristics. In fact, an experienced process operator can usually determine from visual inspection of the chromatogram whether the polymer batch will meet the required specification standards. In many cases a peak molecular weight value, along with the approximate range of molecular weights and the level of unreacted monomers or intermediates can be used as a guide to the quality of the batch. In some instances "good" and "bad" product chromatograms have been generated from which the operator can tell at a glance whether the particular batch needs to be recycled or appears to meet the specifications for the polymer. In some cases, the molecular weight distribution printout at the end of the analysis cycle is used only to confirm the initial observations of the operator and to provide a permanent record for the batch.

B. The Process Size-Exclusion Analyzer

The hardware components used in a process size-exclusion chromatograph have been described elsewhere (Fuller *et al.*, 1979a,b; Mowery, 1980a,b; Roof *et al.*, 1980) and in previous sections; therefore, in this section we shall recall only those major components and features that comprise a process size-exclusion analyzer.

A process size-exclusion analyzer uses a relatively complex sample preparation system to prepare the polymer. A polymer sample preparation system must be able to deliver a representative sample from the reactor vessel or sample point, filter it, and dilute it to an appropriate concentration that does not plug small diameter tubing or degrade the column performance. The dilution chamber as described in Section VIII and Fig. 6 may be used to dilute the polymer to an acceptable concentration prior to its injection. In addition, the sample lines from the reactor to the analyzer's sample-system must be cleaned up after each analysis to prevent polymer build up and contamination of following injections. In some cases, extremely viscous samples are diluted at each reactor using an "on-site" or "satellite" sample preparation system from which only the diluted sample is transported to the analyzer.

The process size-exclusion analyzer also uses a flow controller (Section XI, B), to provide the necessary flow control for meaningful results. Provisions have also been included within the programmer to correct molecular weight values for flow effects based on the retention time of an internal solvent peak. The solvent peak normally has a fixed elution volume at the permeation limit of the column ($K = 1$) and makes an excellent internal marker for correcting minor changes in carrier flow. This technique along with the system's flow control components typically provide a relative standard deviation of 0.1% for the repeatability of a peak's elution time (Fuller *et al.*, 1979b).

Compared with other PLC applications, an MWD requires a much more sophisticated software package for the analysis. In all cases, a microprocessor-based programmer or a computer control system must be employed for an MWD analysis. Some type of digital display or data printing unit is always used as part of the data presentation system in an MWD analysis. Most "size-exclusion programmers" are also normally set up to run periodic molecular weight checks by using molecular weight standards for verification or calibration of the MWD.

The UV, RI, and DC detector are all used for size-exclusion chromatography. The DC detector is especially interesting for an MWD analysis since the response characteristics for the individual constituents of a polymer are usually nearly constant and will often provide an MWD that is more representative of the actual concentration levels than is found with most other LC detectors.

The sample valve is essentially the same valve used with other forms of PLC (see Section XI,D). The major difference is that since macromolecules require substantial dilution, the necessary injection volume is larger (typically 225 μliter).

In summary, process size-exclusion chromatography requires the integration of a number of sophisticated hardware components, whereas other forms of PLC do not necessarily require as sophisticated hardware components for good results. However, in contrast, size-exclusion separations are relatively simple whereas most other forms of PLC require a greater initial effort to obtain a suitable separation for analysis.

XXV. Applications

Actual specific examples of on-line applications are difficult, since such disclosures normally violate the secrecy agreement between the process user and the instrument manufacturer. Sometimes the instrument is purchased as "hardware only," in which case the instrument manufacturer is not necessarily told of the actual application. However, there are a number of application areas that are being serviced by PLC and many more that have the potential of being so serviced. Many of these areas have been discussed elsewhere (Fuller *et al.*, 1979a,b; Mowery and Roof, 1976a,b; Mowery, 1977, 1980a,b, 1981; Roof *et al.*, 1980). Table III also lists some general areas for PLC. There are undoubtedly other industries and streams that could be listed. Also, some on-line problems such as water pollution are common to many industries.

The importance of Table III is that it gives the process engineer and the

TABLE III

PLC APPLICATION AREAS

1. Additives	13. Pharmaceuticals
2. Preservatives	14. Water pollution
3. Stabilizers	(a) Phenols
(a) Diphenyl amine	(b) Petrochemicals
4. Plasticizers	(c) Plasticizers
(a) Dioctyl phthalate	(d) Nitroaromatics
5. Photographic products	(e) Organic acids
(a) Hydroquinone	(f) Polyaromatics
6. Food industry	(g) Chlorinated organics
(a) Flavor components	15. Explosive industry
(b) Preservatives	(a) TNT
(c) Starches	(b) RDX
(d) Heat exchanger fluids	(c) Nitroaromatics
(e) Triglycerides	(d) Nitroplasticizers
(f) Cooking oils	(e) Waste water streams
7. Beverage industries	(f) Nitroglycerin
(a) Carbohydrates	16. Macromolecule industries
(b) Sugars	(a) Polymers and copolymers
(c) Wines	(b) Rubbers and resins
8. Detergents and surfactants	(c) Silicones
(a) Alkylbenzenes	17. Dyes and dye intermediates
(b) Alkylbenzenesulfonates	(a) Anthraquinones
(c) Optical brighters	(b) Benzidine
(d) Water analysis	(c) Naphthols
9. Household chemicals	(d) Water analysis
10. Agricultural products	18. Pigments and formulations
(a) Pesticides	19. Petrochemicals and monomers
(b) Herbicides	(a) Acrylic acid
(c) Insecticides	(b) Acrylic esters
(d) Wood preservatives	(c) Methacrylic acid
(e) Water analysis	(d) Terephthalic acid
11. Organic peroxides	(e) Acrylonitrile
(a) Cumene hydroperoxide	20. Petroleum products
(b) Naphthalene isopropyl hydroperoxide	(a) Water analysis
12. Inorganic ions	(b) Fuels and hydrocarbon analysis

chemist an idea on some of the types of industries and streams that are applicable to PLC. Quite often, whether a specific analysis can be done on-stream depends upon the matrix or components within a given process. In many cases, these components are not even of interest to the process engineer; however, they are part of the matrix from which the process chromatographer must make the separation for the components of interest. Some matrix components also dictate the stability of the column system for PLC. In some applications, a single matrix component within the process determines whether the application is possible by PLC. Furthermore, unless there is a commitment by plant management to provide the necessary training for the service personnel, any PLC application will only survive with difficulty. Thus, Table III is insufficient for determining whether a particular analysis can or should be done by PLC. It is only a guideline that indicates the likelihood that PLC can provide a successful on-stream analysis.

XXVI.   The Future of Process Liquid Chromatography

The use of PLC is expected to expand rapidly during the 1980s, with size-exclusion polymer separations having the potential of becoming the single most common application for PLC. Frost & Sullivan, Inc. (1979) has forecast a 30% annual growth rate for PLC during the next ten years. If this prediction is true, PLC will surpass PGC sales by the year 1990. These predictions are certainly possible, especially if PLC follows the same growth rate trends that are occurring in laboratory LC. Laboratory LC has led over all other laboratory instrumentation in its growth rate for the last eight consecutive years (Thomas and Mosbacher, 1980). Furthermore, it must be recalled that 80% of the known organic compounds (not to mention the inorganic compounds) cannot be analyzed by PGC (Snyder and Kirkland, 1974a). Therefore, one alternative might be PLC.

PLC will not and should not perform all on-line analyses; however, it is a technique that should become familiar to every process chemist and engineer who must monitor and control an industrial process. It is really up to the chemists' and engineers' imagination where and how PLC might serve them. Ultimately, PLC will find its place; its future, however, does indeed appear bright.

ACKNOWLEDGMENTS

The author wishes to express his thanks to the following individuals for their comments and suggestions: Dr. E. N. Fuller, Dr. A. M. Preszler, and Mr. R. K. Bade.

References

Bakalyar, S. R., Bradley, M. P. T., and Honganen, R. (1978). *J. Chromatogr.* **158**, 277–293.
Barrall, E. M., Cantow, M. J. R., and Johnson, J. F. (1968). *J. Appl. Polym. Sci.* **12**, 1373–1377.
Benningfield, L. V., Jr. (1979). *Pittsburgh Conf. Anal. Chem. Appl. Spectrosc., 1979* Paper 123.
Benningfield, L. V., Jr., and Mowery, R. A., Jr. (1981). *J. Chromatogr. Sci.* **19**, 115–123.
Bristow, P. A. (1976). "LC in Practice," pp. 164–165. hetp, Cheshire, U.K.
Broerman, A. B. (1981). *Pittsburgh Conf. Anal. Chem. Appl. Spectrosc., 1981* Paper 854.
Day, D. T. (1897). *Proc. Am. Philos. Soc.* **36**, 112–115.

Dudenbostel, B. F., Jr., and Priestly, W., Jr. (1956). *Ind. Eng. Chem.* **48**(9), 55A–56A.
Frost & Sullivan Inc. (1979). "Industrial On-Stream Process Analyzers," Rep. No. 669, p. 5-1-5. Frost & Sullivan Inc., New York.
Fuller, D. H. (1956). *ISA J.* **3**, 440–444.
Fuller, E. N., Porter, G. T., and Roof, L. B. (1979a). *Pittsburgh Conf. Anal. Chem. Appl. Spectrosc., 1979* Paper 10.
Fuller, E. N., Porter, L. B., and Roof, L. B. (1979b). *J. Chromatogr. Sci.* **17**, 661–665.
Leitch, R. E. (1971). *J. Chromatogr. Sci.* **9**, 531–535.
Littlewood, A. B. (1970). "Gas Chromatography: Principles, Techniques, and Applications," 2nd ed., pp. 129–130. Academic Press, New York.
MacLean, W. (1974). *Am. Lab.* **6**(10), 63–68.
McCoy, R. D. (1978). *Anal. Instrum.* **16**, 155–162.
Martin, A. J. P., and Synge, R. L. M. (1941). *Biochem. J.* **35**, 1358–1368.
Mowery, R. A., Jr. (1977). *Anal. Instrum.* **15**, 119–130.
Mowery, R. A., Jr. (1980a). *Proc. Instrum. Symp. Process Ind., 35th, 1980* pp. 7–17.
Mowery, R. A., Jr. (1980b). *ISA Int. Conf. Proc.* Vol. II, pp. 45–57.
Mowery, R. A., Jr. (1981). *Chem. Eng. (N.Y.)* May 18, 145–152.
Mowery, R. A., Jr., and Roof, L. B. (1976a). *Anal. Instrum.* **14**, 19–23.
Mowery, R. A., Jr., and Roof, L. B. (1976b). *Instrum. Technol.* **23**(6), 43–47.
Munk, M. N. (1971). "Basic Liquid Chromatography," Chapter 5. Varian Aerograph Co., Walnut Creek, California.
National Fire Prevention Association (1977). "Fire Hazard Properties of Flammable Liquids, Gases, Volatile Solids." NFA, Boston, Massachusetts.
National Fire Protection Association (1978). "National Electrical Code," NFPA No. 70-1978, pp. 347–364. NFPA, Boston, Massachusetts.
Parris, N. A. (1978). *J. Chromatogr.* **157**, 161–170.
Porath, J., and Bennich, H. (1962). *Arch. Biochem. Biophys., Suppl.* **1**, 152.
Roof, L. B., Porter, G. T., Fuller, E. N., and Mowery, R. A., Jr. (1980). *In* "Instrumentation and Automation in the Paper, Rubber, Plastics, and Polymerization Industries" (A. Van Cauwenberghe, ed.), pp. 47–53. Pergamon, Oxford.
Sadao, M. (1978). *Anal. Chem.* **50** (12), 1639–1643.
Scott, R. P. W. (1977a). *J. Chromatogr. Libr.* **11**, 69–78.
Scott, R. P. W. (1977b). *J. Chromatogr. Libr.* **11**.
Snyder, L. R., and Kirkland, J. J. (1974a). "Introduction to Modern Liquid Chromatography," 2nd ed., p. 2. Wiley, New York.
Snyder, L. R., and Kirkland, J. J. (1974b). "Introduction to Modern Liquid Chromatography," 2nd ed., p. 126. Wiley, New York.
Thomas, E. J., and Mosbacher, C. J. (1980). *Ind. Res./Dev.* **22**(2), 160–164.
Walker, S. E., Mowery, R. A., Jr., and Bade, R. K. (1980). *J. Chromatogr. Sci.* **18**, 639–649.

6

Automation in the Clinical Chemistry Laboratory. I. Concepts

CARL C. GARBER

Clinical Laboratories and Department of Pathology and Laboratory Medicine
University of Wisconsin-Madison
Madison, Wisconsin

and

R. NEILL CAREY

Clinical Laboratories
Peninsula General Hospital Medical Center
Salisbury, Maryland

I. Introduction

Computers were first applied to the clinical laboratory during the mid-to-late 1960s. They were first used as on-line data loggers, monitoring the analog outputs of the Autoanalyzers (Technicon Instruments Corporation, Tarrytown, New York) by means of retransmitting slidewires on the Autoanalyzer recorders. The analyzers' operation, even for the six- and twelve-channel analyzers, was programmed by setting switches and placing pins on program boards in the analyzers. During this period, there came the realization that the computer could control an analyzer's functions in addition to monitoring the analytical data. The first commercial clinical analyzers with integral computerized process controllers appeared around 1970. By 1980, nearly every analyzer coming onto the market was microprocessor-controlled.

There were several pressures that led to the rapid proliferation of computer- or microprocessor-controlled equipment. The first was the phenomenal growth of volume of clinical chemistry testing. Between 1970 and 1980 this volume grew at a rate of 10–20% per year in many laboratories. Higher throughput analyzers were essential to meet this growth efficiently. Throughput of the available technologies could not be increased without computerization if accuracy was to be maintained and improved. Many instrument concepts, for example, centrifugal analyzers, were simply not feasible without total computerization. Also, there were signs that the available analyzers were sometimes run under inadequate quality control systems; computer intervention was seen as a means for improvement. It became necessary to perform a wider variety of tests in high volume. The complexity of the analyzers grew. For example, there were recommendations that enzymes be measured by true rate methods using multipoint detection. Another area of pressure was the increasing volume, variety, and complexity of STAT analyses (performed rapidly, for life-threatening conditions of the patient). STAT testing could no longer be performed efficiently by manual procedures alone. A whole new category of automated analyzers developed, not so much to address the problem of high volume but to provide a variety of procedures quickly with a minimal set-up.

Finally, and very significantly, microprocessors became very inexpensive. Even with all the reasons for the growth of automation given above, there would not have been as much if the price of process control computers had not fallen. Automation has made laboratory testing a relative bargain in the health-care delivery system; thus, laboratory testing has been utilized increasingly.

On the other hand, there are growing pressures and incentives to reduce the volume of laboratory testing under the guise of reducing total health-care costs. Users are increasingly concerned about the cost-effectiveness of automated analyzers. In the beginning, hospital laboratories sometimes acquired automated instruments without careful cost analysis. With increasing federal control, and with several states adopting strong rate-review mechanisms, capital equipment funding is becoming scarce. In some hospitals in Maryland, automated analyzers are on a 10-yr depreciation schedule dictated by the Health Services Cost Review Commission. The user must be very careful to select analyzers that will have long useful lives.

The intent of this chapter is to describe the total analytical process encountered in the clinical laboratory and to describe features of an ideal or state-of-the-art analytical system. The following chapter describes several commercially available automated systems with process control, where the process extends from the request for testing to the receipt of the test results. The chapter deals with those analytical systems specifically designed and oriented toward clinical chemistry rather than general purpose analytical systems. These special purpose systems best illustrate the incorporation of process control. The discussion is directed toward analytical chemists who are not actively working in clinical chemistry.

II.   Automation and Process Control: Definition of Terms

The concepts of automation and mechanization have fundamentally different connotations as they relate to process control, yet they are often used interchangeably in the clinical laboratory setting. The Commission on Analytical Nomenclature of the International Union of Pure and Applied Chemistry has recommended specific definitions for automation and mechanization (Stockwell and Foreman, 1979). The distinctive characteristic in automation is the concept of a feedback mechanism to control at least one operation without human intervention. Feedback control is absent in the strict meaning of mechanization.

III.   The Analytical Process in Clinical Laboratory Testing

If an automated analyzer is to be maximally useful, it must be compatible with the flow of test requests and data through the laboratory. Many large hospital laboratories and commercial laboratories have central laboratory computer systems and/or hospital information systems (HIS, a central hospital-wide computing system). Most smaller laboratories handle test requisitioning and data reporting manually. The sequence of events in testing for the manual extreme and the computerized extreme are shown in Fig. 1. The left side shows the work flow in all laboratories prior to the 1960s. The right side shows that it is now technically possible to eliminate human transcription of requests or data except for the original request for a test, and the identification of the patient when the specimen is obtained. The workflow in clinical chemical analysis has been described in detail by de Haan (1979). Certainly, the 1980s will see a continuation of efforts to integrate analytical systems into the total HIS to take advantage of the relative speed and freedom from human errors offered by computerization. In smaller laboratories, there will be a continued replacement of individual steps in the manual workflow by automation.

IV.   General Features of an Ideal, State-of-the-Art Analyzer

The incorporation of the microprocessor into an automated analyzer makes many new functions possible. These improvements are being realized in the areas of instrument control and data acquisition. More significantly, the micro-

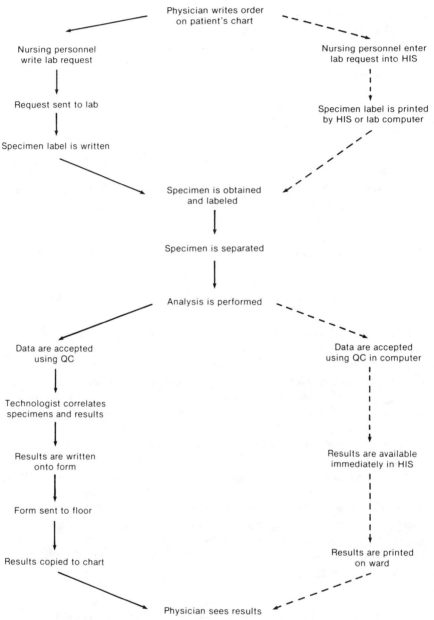

Fig. 1.   Analytical process in the clinical laboratory for manual laboratory (→) and computerized laboratory (⋯→).

processor enables the extension of the analytical process to include on-line data review and management. To serve as a basis for the discussion following on the present state of the art of automated systems and to provide impetus in the development and direction of new generations of automated systems, we wish to discuss the philosophy and the characteristics of the ideal high-volume automated clinical analyzer using present day technology. Present automated analyzers have sample throughputs in the range of 100–200 samples per hour; humans struggle to keep up with samples and data. Most state-of-the-art analyzers are precise and accurate enough to meet clinical needs for the commonly tested analytes (Ross *et al.*, 1980). Additional consolidation of the laboratory services into one major multichannel analyzer may not be beneficial. Any mechanical or electronic failures that cause that single large instrument to be nonoperational will have a major impact on laboratory operation unless there is a backup system. Thus, our comments about the ideal analyzer will be a description of instrumental features with regard to the total analytical process, rather than a description of precision and throughput.

Automated analyzers have previously been viewed as mechanical devices into which one fed specimens and from which one received results of analyses on those specimens. Microprocessors have enabled dramatic improvements in automated analyzers because control of automation can now be programmed, with relatively easy program changes, and because the processor can store and manipulate data. It is these abilities of the computer that can be built upon to produce analyzers that are easier to operate, more reliable, more complete in scope, and more compatible with the use of computers.

The microprocessor offers improved communication between the operator and the increasingly sophisticated analyzer. The analyzer should in turn enable the operator to program the performance of the various functions of the analyzer in order to perform analyses. This should be the case for any general-purpose clinical chemistry analyzer. User-programming of chemical experiments employing a minicomputer was demonstrated first with LABTRAN (Toren et al., 1972) and then with WISDOCS (Toren *et al.*, 1973a,b). The user should have access to the microprocessor through a fully documented computer programming language, making it possible, if necessary, to document, troubleshoot, or modify the operating parameters of the system. Modifications of analyzers by users have in the past promoted the development and marketing of improvements to existing instruments and of new instrument designs.

Evenson (1979) has stated that operator–microprocessor communication needs to be standardized. At present, every microprocessor-controlled analyzer has its unique instruction set and information/error message language. Often the messages consist of only one or two letters having different meanings for each analyzer. Standardized coding will become more important with greater utilization of microprocessor-controlled systems in the laboratory. Standardized coding and keyboard systems should serve as a basis for development and need not inhibit further sophistication of operator–microprocessor communication.

The microprocessor assists the user in monitoring instrument function at many critical stages. Already most instruments can alert the operator when the incubation temperature is out of control or a conveyor is jammed. This could be refined toward self-diagnosis of problems by the analyzer. For example, a microprocessor with sensors at strategic locations in an analyzer could communicate not only that the conveyor is jammed, but also the general area of the jam. Some of the troubleshooting algorithms now used by humans can be programmed without costly, additional sensors. For example, the instrument could alert the operator that the photometer light source is weak by monitoring photocurrent, as well as absorbance.

Another key improvement offered by the microprocessor is its applicability to specimen identification. This is still an area in which human error might very well ruin an otherwise acceptable analysis; in fact, this source is quite likely one of the larger sources of error in most laboratories. In an ideal system, the patient would have a machine- and human-readable identification label on the wristband identification bracelet. The technologist obtaining a blood or urine specimen from the patient would carry a device to "read" the wristband I.D. and produce machine- and human-readable labels to be attached to the specimen at the bedside. These labels should also identify the specimen by a "reader"-specific accession number and indicate the time when the specimen was obtained. Upon receipt of the specimen and request in the laboratory, a computer input device would scan the request card and label and automatically enter the information into its files. As the specimen travels through the laboratory, each analyzer would recognize the specimen. Analyzers would accept specimens in random order, perform the analysis, print out results with identification number, and transmit acceptable sample analysis data with sample identification to the central laboratory computer. Finally, the central computer would collate all test data for the patient. Most of these steps are now available individually. They need only be interfaced to each other so that manufacturers and users can develop flexible systems providing total specimen identification throughout.

Ideally, the analyzer should sample the specimen from the container in which it was collected. This reduces the labor expense, transcription errors, and exposure of technologists to disease that accompany manual transfers of specimens to sample cups. Blood specimens are usually centrifuged prior to analysis to separate the serum supernate from the red cells. A firm physical barrier between serum and clot can be obtained by use of silicon polymers in special blood collection tubes or by insertion of a press fitting filter. With the silicon polymer separator, the specimens can be stored without degradation until analysis (Laessig *et al.,* 1976), and the analyzer can sample the specimen from the original collection container.

If the specimen must be transferred to another container prior to analysis, duplicates of the original label should be transferred also, and the analyzer should be able to read the duplicate label.

Specimens are placed into racks or turntables to await their turn to be sampled. Specimen identification by the sampler should not take place until the time of sampling, to permit emergency specimens to be inserted with high priority. Any turntable malfunctions should be communicated to the user.

After the specimen is sampled, completeness of sampling and sample introduction should be verified by confirming that the required volume of sample was obtained and delivered, and that no clots were present. Most analyzers now use digital pipets under program control. The sample volume programming should be modifiable by users for special applications such as pediatric specimens.

The microprocessor also offers flexibility for reagent volumes, timing of reagent addition, and length of incubation. For some procedures, the reagent is not added until immediately before detection, whereas for others, long incubations are necessary. Often, a long preincubation is followed by initiation of a measurement reaction immediately before detection. Again, users should have programming flexibility to enable adaptation of a wide variety of types of methodologies to the analyzer.

Incubation temperature should be monitored, and error signals should accompany the results of any analysis performed during the error condition. Many analyzers identify only those temperature problems that occur during the readout stage.

The ideal analyzer should provide flexibility for the user to define the parameters that control the data acquisition process. The time for starting the measuring process, the number of measurements averaged per data point, the time interval between data points, and the number of data points should be definable by the user in accordance with the principles of the method and the characteristics of the instrument and detector. The detection system should be monitored frequently or even continually to correct for drift in the detection system, whether it be the light source or the detector itself.

The ideal analyzer should offer the user a wide selection of calibration approaches. For linear methods, the techniques should include (1) the use of theoretically derived conversion factors to convert the raw measurement into test units, (2) one-point calibration with suitable blanking, (3) two-point calibration with suitable blanking, and (4) multipoint calibration with suitable blanking utilizing linear regression techniques with limit checks to determine the calibration line of best fit. Nonlinear methods for immunochemical assays may require multipoint calibrations utilizing point-to-point, log-logit, spline, high-order polynomial regressions, and power function regressions (including log functions) to describe the calibration curve of best fit.

A human operator cannot keep track of results of testing on quality control specimens since they are analyzed on a high-speed multichannel analyzer without some computer help. Quick *et al.* (1980) have described a quality control monitor for use with the Technicon SMAC (3000 data/hour). It shows the operator a visual graphic quality control report on each channel in real time on a CRT. Quality control problems, caused by violations of various control rules, produce different colors and shapes, thereby informing the operator which channels have problems and how bad they are. Ideally, quality control monitors should combine results from control specimens with analyzer malfunction diagnostics to help the operator determine whether the problem is chemical or instrumental.

Reports of results should include specimen accession number, cup number (tray location), test name, and result. Abnormal results should be so identified. The format of the printout should be flexible in order to meet the needs of different laboratories internally, as well as the various external users, whether physicians, clinics, or other hospitals.

For rapid integration of automated analyzers into the hospital information systems, there should be standardization, or at least flexibility, in the format of data streams coming from the analyzer. The data stream should have a flexible baud rate. Most automated systems are RS-232 C compatible. There should be a universal software interface in the central computer so that special programming (which is expensive) is not necessary to interface new instruments. It should only be necessary to inform the computer of the format and timing of the instrument's data stream and plug in the analyzer to prepare the computer and analyzer for operation together. A. A. Eggert (private communication, 1981) has recommended a format for the data stream from the analyzer to the computer. The data stream should be standard 8-bit ASCII of odd parity. All transmissions should start with an STX (002) and end with an ETX (203). Each line of the transmission should contain 20 characters. The first 4 identify the item (left justified), the next 2 are spaces, the next 6 are the numeric result (right justified), the next 2 are spaces, the next 4 are flags (left justified) and the last 2 are for carriage return/line feed. This is illustrated as

IIII- -NNNNNN- -FFFFCL

The item identifier, I, should be a code such as LDH, GLUC, or a cup number. The numeric field, N, should contain a decimal point as necessary and should contain dashes (- - - - - -) if a result was requested, but no answer could be obtained. The flag field, F, should consist of four one-character flags, which can be letters, numbers, or punctuation. It can be used for informational purposes (i.e., to distinguish between patient samples and controls) or to indicate instrumental problems. The coding for items and flags should be guided by common usage (i.e., NA for sodium or > for exceeded range). The first item transmitted should be the cup number and the second, if known, the specimen number. After this the results should follow. Each specimen should produce a separate transmission.

The central computer (either laboratory or HIS) should be interfaced to instruct the automated analyzer to perform only the tests actually requested on individual samples. One of the advantages of discrete multichannel analyzers is that only the tests desired are performed, so that reagents are not used needlessly. Usually, an operator must inform the analyzer by the CRT keyboard or marked sense cards which tests are requested on a particular specimen. There is a savings in labor if the dual manual-requesting of tests is reduced to a single request into HIS. This also has implications for on-line quality control monitoring. In some laboratories with a central computer, all tests available on the analyzer are run, and data are transmitted to the computer. The computer ignores the unrequested test results. If a channel goes out of control, the quality control monitor on the analyzer has no way of knowing whether the out-of-control test was really ordered on any individual patient specimen. Thus, specimens are tested again, even though the test was not requested.

Two-way communication would have the effect of making data available to the nursing personnel and the physician on the ward through HIS more rapidly. The data would be available as soon as they were verified by the instrument operator, without duplication of effort or needless repetition of tests.

Time records are very important when questions arise about individual specimens. The computer should maintain records of the times when the test was requested, when the specimen was obtained, when the analysis was performed, and when the results were approved and reported. If the complete analytical process is conducted under the control of the laboratory computer or HIS as described here, this documentation can be tabulated automatically.

V.   Conclusion

Although the emphasis of these chapters is a description of automated instrumentation, it should be recognized that the finest instrumentation by itself will not necessarily produce medically acceptable results. The selection and use of any instrument should be based on (1) the practical features of the instrument, (2) the principles of the analytical methods, (3) the quality of reagents, (4) the acceptability of analytic errors relative to medical requirements, and (5) the availability of a sensitive quality control program.

Westgard et al. (1978) have described the basic concepts for the selection of methods and for decisions about the acceptability of their analytic performance. Obviously, the method (instrument and reagents) must be capable of meeting the practical needs of the laboratory in terms of sample size, sample throughput, turnaround time, cost, etc. The chemical reactions and instrumental measurements of the method must be sensitive and specific for the analyte. Ideally, they should be similar to those of the concensus reference method for that analyte.

The analytic errors of a method should be measured by an experimental design and statistical treatment that ensures reliable estimates of the magnitude of specific errors. Random error describes the reproducibility of results. Systematic error describes the agreement of results reported by the method to the analyte concentration reported by an accepted reference method. Since interpretation of systematic error is so dependent on the quality of the method used as a reference method, only methods that are of the highest quality and are nationally recognized should be selected as reference methods for comparison. To decide objectively whether errors are acceptably small, experimentally determined estimates of error must be compared to medically allowable errors. The medically allowable error for an analysis is the maximum error that will not adversely affect the patient care strategy. Statistical limits on the estimates of error are suggested (Westgard *et al.*, 1974) to enable the investigator to judge errors with some defined confidence (for example, 95% limits).

The National Committee for Clinical Laboratory Standards (NCCLS) has proposed a guideline (NCCLS, 1976) for manufacturers to follow to develop statistically valid claims for precision and accuracy. This document has been revised and expanded into three documents on concepts and preliminary studies (NCCLS, 1979, EP2-P), precision (NCCLS, 1979, EP3-P), and accuracy (NCCLS, 1979, EP4-P). More recently, the authors have participated in an NCCLS subcommittee effort to develop guidelines for product users to evaluate system performance relative to performance claims or criteria of medical usefulness (NCCLS, 1982). After the instrument's performance has been shown to be acceptable, the user must establish a quality control program to ensure that performance remains acceptable.

Traditionally, quality control in the clinical laboratory has been based on ±2 standard deviation control limits. Although this quality control procedure is sensitive to errors, it also has a high frequency of false rejections (the result is outside statistical limits for statistical reasons when no error exists). This is particularly true for multichannel analyzers. False rejections increase dramatically as the number of observations or analytical channels increases when a ± 2 standard deviation quality control limit is used (Westgard *et al.*, 1977). Refined multirule quality control programs should be implemented to minimize the false rejections yet maintain sensitivity to real errors. Furthermore, the multirule quality control system should indicate whether the out-of-range condition is due to an increase in random error or systematic error. Westgard and Groth (1981) have developed a procedure for defining a quality control system that has a known probability of detecting a predetermined amount of error with a known probability of false rejection. Application of these advanced quality control systems will enable users to be confident that their instruments are operating within medically acceptable error limits.

References

de Haan, J. B. (1979). *In* "Topics in Automatic Chemical Analysis" (J. K. Foreman and P. B. Stockwell, eds.), Vol. I, pp. 208–236. Wiley, New York.

Evenson, M. E. (1979). *Anal. Chem.* **51**, 1411A–1413A.

Laessig, R. H., Westgard, J. O., Carey, R. N., Hassemer, D., Schwartz, T., and Feldbruegge, D. (1976). *Amer. J. Clin. Pathol.* **66**, 653–657.

National Committee for Clinical Laboratory Standards (1976). "Protocol for Establishing the Precision and Accuracy of Automated Analytic Systems," NCCLS, PSEP-1. NCCLS, Villanova, Pennsylvania.

National Committee for Clinical Laboratory Standards (1979). "Protocol for Establishing Claims for Clinical Chemical Methods: Introduction and Performance Check Experiment, Replications, and Comparison of Methods Experiment," NCCLS, EP2-P, EP3-P, and EP4-P. NCCLS, Villanova, Pennsylvania.

National Comittee for Clinical Laboratory Standards (1982). "Proposed Guidelines for User Evaluation of Precision Performance of Clinical Chemistry Devices," EP5-P. NCCLS, Villanova, Pennsylvania.

Quick, R. F., Thew, C. A., and Thiers, R. E. (1980). *Clin. Chem.* **26**, 1014.

Ross, J. W., Martin, D. F., and Moore, T. D. (1980). *Am. J. Clin. Pathol.* **74**, 521–530.

Stockwell, P. B. and Foreman, J. K. (1979). *Top. Autom. Chem. Anal.* **1**, 14–43.

Toren, E. C., Jr., Carey, R. N., Sherry, A. E., and Davis, J. E. (1972). *Anal. Chem.* **44**, 339–343.

Toren E. C., Jr., Carey, R. N., Cembrowski, G. S., and Shirmer, J. A. (1973a). *Clin. Chem.* **19**, 1114–1121.

Toren, E. C., Jr., Mohr, S. A., Busby, M. G., and Cembrowski, G. S. (1973b). *Clin. Chem.* **19**, 1122–1127.

Westgard, J. O., and Groth, T. (1981). *In* "Proceedings of the VIIIth Tenuvus Workshop, Quality Control in Clinical Endocrinology" (Wilson, D. W., Gashell, S. J., and Keag, K. G., ed.). Alpha Omega Publ., Cardiff, England (in press).

Westgard, J. O., Carey, R. N., and Wold, S. (1974). *Clin. Chem.* **20**, 825–833.

Westgard, J. O., Groth, T., Aronsson, T., Falk, H., and de Verdier, C. (1977). *Clin. Chem.* **23**, 1857–1867.

Westgard, J. O., de Vos, D. J., Hunt, M. R., Quam, E. F., Carey, R. N., and Garber, C. C. (1978). "Method Evaluation." Am. Soc. Med. Technol., Bellaire, Texas.

7

Automation in the Clinical Chemistry Laboratory. II. Classification and Examples

CARL C. GARBER

Clinical Laboratories and Department of Pathology and Laboratory Medicine
University of Wisconsin-Madison
Madison, Wisconsin

and

R. NEILL CAREY

Clinical Laboratories
Peninsula General Hospital Medical Center
Salisbury, Maryland

I. Classification of Process-Controlled Analyzers in Clinical Chemistry

Before proceeding with a discussion and review of process control in the clinical chemistry laboratory, it is desirable to define a basis of classification of analyzers that is logical and fundamental. There is no single classification system of automated instruments that is generally accepted (Schwartz, 1978; de Haan, 1979; Mitchell, 1980). The terms "continuous flow" and "discrete" analyses are often used to discriminate between fundamental characteristics. These terms, in themselves, are general and not precise. To further define the characteristics of an automated system, other terms such as "single channel" or "multichannel" are used. The latter in particular is ambiguous because of its loose application to all systems providing two or more analytical outputs per sample regardless of the instrumental approach. Some intermediate configurational information is necessary to classify automated systems more specifically.

The term "continuous flow" has traditionally had an unambiguous meaning, namely, the process described by Skeggs (1957), whereby air bubbles are inserted into the continuously flowing analytical stream. But the recent development of the unsegmented continuous flow technique (Margoshes, 1977; Betteridge, 1978; Ranger, 1981) now results in possible ambiguity of meaning. To eliminate confusion, the general term "continuous flow" can be subdivided into "segmented flow analysis" (SFA) and "flow-injection analysis" (FIA) as proposed by Ranger (1981). Segmented flow analysis and flow-injection analysis are discussed separately in this chapter with illustrative examples.*

The other major approach to processing the sample–reagent mixture is by discrete containers. These may take the form of traditional open test tubes that are transported by means of a carrier rack or carousel, or the form of separated cuvets in centrifugal analyzers, or specialized containers (plastic bag in the duPont aca), or specialized thin films (Kodak Ektachem). A concept intermediate to the continuous flow and discrete systems is the in situ analytical system. Although at first thought this technique seems to be discrete, it may also be viewed as a "stopped flow" example and in fact may be subject to sample carryover just as in continuous flow techniques. Each of these types or classes is discussed here with examples of process control.

The terms "single-channel" or "multichannel" analyzer, when superimposed on these basic classifications, produce a less ambiguous description of automated analysis (Table I). For example, for the segmented flow analysis, there will be (1) the dedicated single-channel analyzer such as the traditional AutoAnalyzer (Technicon Instruments Corporation, Tarrytown, New York), (2) batchable single-channel analyzers, which require changing the tubing manifold

---

*Segmented flow and flow injection are also discussed in Chapter 2.

TABLE I

Classification of Automated Analysis

Number of tests	General technique Specific technique	Continuous flow			Discrete analysis		
		SFA	FIA	in situ	Test tube	Centrifugal analyzer	Prepackaged
Single		X	X	X	X	X	X
Multibatch		X		X	X	X	
Multiparallel		X		X	X	X	X
Multisequential					X		X

between test methods, and (3) the multichannel segmented flow analyzer, involving a variety of different channels operated simultaneously in a parallel mode. FIA has only been performed on a single-channel basis. The specifically designed stainless steel manifold is not conducive to changes for different methods.

In the discrete analyzer classification, the open tube subclass can be operated (1) as a dedicated single-channel analyzer, processing samples for the same test in sequence, (2) as a batchable single-channel analyzer requiring changes in reagents and instrumental parameters between batches of different tests, (3) as a single-channel, variable test analyzer, i.e., different tests can be analyzed automatically in sequence, or (4) as a multichannel analyzer in which different tests are assayed concurrently or in parallel for the same sample.

The centrifugal analyzer offers a unique classification of automated analysis, namely, multichannel, single-test analysis in parallel with batching capability to change the test type. Recently, software has been developed to make it possible to analyze different tests in the same run.

The pack or plastic bag discrete analyzer offers only single-channel analysis of variable tests in sequence. The thin film discrete analyzer has been developed along two lines, the multichannel, parallel analyzer and the single-channel analyzer with sequentially variable tests.

The final classification of automated analyzers is the discrete in situ analyzer. These can be operated as (1) dedicated single-channel analyzers, (2) batchable single-channel analyzers requiring changes in reagents between batches of different tests, or (3) as multichannel analyzers operating in parallel.

Thus, the classification system that serves as a basis for this outline keys on specific techniques in the general categories of "continuous flow" and "discrete" analysis. These specific techniques are subdivided by considering the various configurations of single- and multichannel concepts. The remainder of this chapter is a description of characteristics and examples of the different classes of analyzers defined previously. The specific examples discussed are not necessarily felt by the authors to be better than others. The example analyzers were chosen because of their popularity, because of the authors' experience with them, or because of their new technology.

II. Continuous Flow Analysis

A. *Segmented Flow Analysis*

1. BACKGROUND

Segmented flow analysis dominated the mechanization and automation efforts in the clinical laboratory for at least ten years following a report of its development by Skeggs (1957).

It was not until these ten years of rapid application of SFA had elapsed that descriptions of SFA on a theoretical basis were reported. Thiers and co-workers provided the initial discussion of the kinetic features of SFA (Thiers *et al.*, 1967), the effect of bubble gating in the flow cell (Habig *et al.*, 1969), an analysis of peak characteristics (Evenson *et al.*, 1970), and an analysis of the lag phase in SFA (Thiers *et al.*, 1971). Walker and co-workers (1971; Walker and Andrew, 1974; Walker, 1976, 1977) also discussed theoretical aspects of SFA. Snyder and Alder (1976a,b) and Snyder (1976, 1977) presented fundamental discussions on the factors contributing to dispersion in SFA. Spencer (1976) summarized the factors influencing the observed characteristics of SFA. These and many other reports contradict the comments by de Haan (1979) that a theoretical basis for SFA was lacking. As the theoretical basis of SFA was being developed, refinements in the measuring electronics (Thiers *et al.*, 1970; Carlyle *et al.*, 1973; Neeley *et al.*, 1974) were reported.

The increase in the understanding of the principles of the SFA technique is demonstrated in the developmental trail of segmented flow systems by Technicon Instruments Corporation, which began with the AutoAnalyzer in 1957, the AutoAnalyzer II in 1960, the SMA 12 (30 samples per hour) in 1965, the SMA 12/60 and 6/60 in 1967, the SMAC (150 samples per hour) in 1972, the SMA II 12/90 in 1976, and most recently, the SMAC II in 1981.

The early systems were mechanized units with a passive monitor, the analytical signal detector, which displayed the analytical signal on a strip chart recorder or storage oscilloscope (for the SMA 12/60 and 6/60). In the multichannel SMA 12's and SMA 6's, phasing between channels was achieved with respect to the slowest channel by inserting a delay coil of appropriate length into the flowing stream for each channel. Thus, the input of a single pen chart recorder was automatically switched from one detector to the next to display each channel's analytical signal in sequence for one particular sample. Each day it was necessary to reoptimize the interchannel phasing. It was also incumbent upon the operator to identify sampling or hydraulics problems from the shape of the response curves. Selectivity of tests is achieved only by suppressing data. All methods are run on each sample processed.

2. HIGH-SPEED COMPUTER-CONTROLLED SEGMENTED FLOW ANALYSIS

A quantum jump in segmented flow systems was realized with the advent of the SMAC (sequential multiple analyzer plus computer) in 1972. This was Technicon's first attempt to utilize a computer to control and monitor the various functions of sample indentification, sampling, hydraulics, detection, data manipulation, and data reporting. Details of the performance and operation of the process-controlled SMAC were presented in reviews by Schwartz *et al.* (1974; Schwartz, 1978), Westgard *et al.* (1976), Karcher and Foreback (1977), Rush and Nabb (1977), Robertson and Young (1977), and Carey *et al.* (1977). The computer-controlled system eliminated the necessity for phasing the channels relative to one another. Each channel was monitored independently by the computer, and the analytical data was stored in the data buffer in sequence and collated for printout. Autocalibration is performed using set points stored in computer memory from the program tape. The hydraulics were refined to allow analysis of 150 samples per hour. A high-precision, electromechanical sampling system was controlled by the computer. The analytical stream was miniaturized. Insertion of air bubbles into the analytical stream was controlled, and the design permitted the air bubbles to pass through the flow detector. The computer was able to detect the presence of air bubbles in the flow cell by discrimination against absorbances above a certain threshold. There are air bubbles in the flow cell about two-thirds of the time.

A fundamental component of the software in the dedicated process control computer was the peak shape monitoring system. Percy-Robb and co-workers (1978) have studied the performance of this system. Successful operation of the SMAC depended on the sensitivity of this software in detecting process errors, whatever they might be, and then alerting the operator of their presence. Robertson *et al.* (1979) studied the magnitude of errors for results that have been flagged by the peak monitor and found that although some flagged results were correct, a significantly high proportion of flagged results were incorrect and thus should not be reported.

A number of modifications of the SMAC have been reported. Errors due to air bubble compression in the sample line have been reported by Walmsley *et al.* (1980). Reduction of sample volume was discussed by Stamper and Robertshaw (1980). Manifold changes were made in certain channels to improve operational performance (O'Leary and Duggan, 1980; Garber *et al.*, 1981). Other improvements in the overall analytical performance have been made by using immo-bilized enzymes to achieve more specific chemical analyses and a reduction in reagent costs (Horvath and Pedersen, 1977; Leon et al., 1977; Garber et al., 1978).

With the advent of a high-speed multichannel analyzer, such as SMAC, has come a serious problem, that of monitoring the data spewing out at the rate of 3000 results per hour. Even if the microprocessor can flag results affected by process errors, the corresponding specimens must be identified and saved for reanalysis after the problem is resolved. A more difficult task is to monitor the quality of the results by use of a quality control program, using control speci-mens to identify the analytical errors over and above the system-identified pro-cess errors. If a quality control material is tested in every tenth cup, for example, there are 300 quality control results that must be reviewed each hour by the operator. An on-line system of computer-assisted review of the quality control data and computer-assisted editing of the patient data has been reported by Eggert and Westgard (1975). Control limits are stored in memory for each test for each control material. Specific control materials are identified by uniquely assigned accession numbers; thus, the computer is able to determine whether the results are within acceptable limits and notify the operator. More sophisticated applications of different quality control rules and limits have been developed by Westgard et al. (1977a,b, 1979; Westgard and Groth, 1979) to improve the sensitivity of the quality control program in detection of analytic errors and to minimize the false rejections. An on-line monitor using this approach has been described by Quick et al. (1980).

SMAC II (introduced in 1981) operates with the same basic characteristics as the original SMAC, i.e., 150 samples per hour, 20 channels, and a peak shape monitoring system. However, there are several major changes at the front end of the analyzer and in the data handling facilities (J. Levine, private communica-tion, 1981).

The positive sample identification system has been changed in format (bar code) to give more reliable identification. The electromechanical sampler has been changed to a microprocessor-controlled pneumatic system that operates on the same sampling sequence as its predecessor. A special STAT feature has been incorporated to allow priority sampling and priority reporting of emergency specimens. A key refinement in the hydraulics-monitoring capabilities is an air bubble detector in the predilution cartridge. This detector primarily automates the LOAI (liquid-out, air-in) adjustment on each sample. This same bubble detector is able to identify clot problems as well as short sample conditions. These fundamental improvements in the monitoring of process errors supplement the peak shape monitoring capabilities. In the original SMAC, a number of different hydraulics errors were fatal and caused the system to halt or stop. In SMAC II, the detected errors are displayed on a CRT, and the operator can interact with the computer to correct the fault without bringing the SMAC down. If one channel has difficulty, the operator can instruct the computer to continue with the analyses of all the other channels.

The computer is able to store the acceptable portions of a sample profile in temporary buffer, to be combined with the results of tests that are repeated because of a hydraulics problem or a quality control problem for a particular test. The computer is also able to store information for each test that describes the expected or reference range of a particular analyte, those limits beyond which it is important to review all data, and critical values beyond which life may be threatened. In the latter case the operator is immediately notified so that the physician may in turn be notified. The software for the system has been expanded to store statistical data for up to 16 different quality control materials. The computer performs on-line multirule review of the quality control data.

3. AUTOMATED RADIOIMMUNOASSAY BY SEGMENTED FLOW ANALYSIS

In radioimmunoassay (RIA), the analyte is the antigen in an antigen–antibody reaction. There is competition for a limited number of antibody sites between a fixed concentration of radioactively labeled antigen and the unlabeled antigen in the sample (Yalow and Berson, 1959). The antibody (and with it the bound radio-labeled antigen and bound unlabeled antigen) is separated from the unbound antigen and counted. Alternatively, the free radio-labeled antigen can be counted. Then a standard curve of antigen concentration versus bound (or free) radioactivity is constructed, and antigen concentrations in patient specimens are quantitated by interpolation. Relatively low test volumes and technical difficulties in automating the separation of free and bound radioactivity have minimized the impact of automation upon radioimmunoassay in most hospital laboratories in the past.

Several automated RIA systems have been developed commercially. The segmented flow approach has been adopted in the Gammaflo (E. R. Squibb & Sons, Inc., Princeton, New Jersey) and the STAR (Technicon Instruments Corporation, Tarrytown, New York 10591). Both of these analyzers are computer controlled, and both rely on a system of valves operating in sequence to direct the antibody-bound antigen to the counting chamber and to direct the free antigen to waste. The major difference is in the technique used for separating the antibody-bound antigen from the free antigen.

In the Gammaflo, described by Brooker *et al.* (1976) and Bowie (1980), separation is effected by directing the sample–antibody–radio-labeled antigen incubation mixture through an ion-exchange column, thus trapping the free antigen. The antibody-bound radio-labeled antigen is eluted from the column into the counting chamber by a buffer solution. Valve actions redirect the flow, and the column is stripped by a wash solution, while the bound radio-labeled antigen is being counted. Then the sample is evacuated from the counting chamber; the chamber is rinsed; and the column is prepared for the next specimen. Data reduction is very flexible; spline, third-order polynomial, log, or log-logit computational methods can be chosen (E. R. Squibb & Sons, Inc., 1980). Performance of the Gammaflo has been evaluated for digoxin and cortisol by Valdes *et al.* (1979). Adaptation of a procedure for RIA of urinary cyclic AMP to the Gammaflo at a sampling rate of 60 samples per hour has been reported by Brooker and Murad (1980).

In the Technicon STAR (Cohen and Stern, 1977; Bowie, 1980), automated separation of the bound and free antigen is enabled by covalently bonding the antibody molecules to an organic polymer (alpha-cellulose), in which ferric oxide particles are also encased (Chen, 1980a). A magnetic field is operated under computer control to trap the solid-phase antibody while the free antigen is washed to waste. When the magnetic field is switched off, the solid phase antibody particles wash through a flow-through gamma counter, where the antibody-bound radio-labeled antigen is counted.

In the version of STAR being marketed (Technicon Instruments Corp., 1980) the hydraulic system functions under the control of a microprocessor, somewhat analagously to SMAC. A simplified flow diagram is shown in Fig. 1. Antibody reagent and radio-labeled antigen are added to the analytical stream only when the instrument senses the presence of sample; at other times buffer solution is pumped and recycled to maintain constant flow rates and pressures.

The sampler operates under computer pneumatic control, aspirating 100-μliter samples at the rate of 60 per hour. Multiple aspirations can be performed from the same cup; each specimen can be analyzed once, twice, or four times with a corresponding decrease in throughput. A light-activated switch, the "sample error detector assembly," monitors sample flow for phasing purposes and detects such error conditions as absence of sample, short sample, long sample, or clogged lines.

Radio-labeled antigen is connected to the analytical stream by a hydraulic circuit that returns the reagent to its container when the sample is not present. When the radio-labeled antigen is required, pinch valve 5 is activated to let it flow into the analytical stream. A light-activated switch monitors phasing and flow of this reagent.

The solid-phase antibody is also recycled to a storage reservoir when it is not required. There it is continually stirred to maintain it in suspension. Pinch valve 4 is activated to let it flow into the analytical stream at the "mix tube" when the sample is present. The flow of this reagent is also checked by the "solid-phase error detector assembly."

The sample, radio-labeled antigen, and solid-phase antibody are mixed and spend 10 min flowing through a 37°C incubation coil. Thus all STAR assays

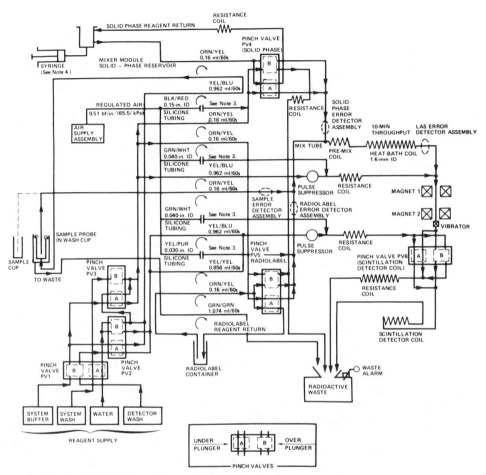

Fig. 1. Simplified flow diagram of the STAR system.

Notes:

1. Pinch valves PV1, PV2, and PV3 are spring returned. With system power off, the "B" side is closed.

2. Pinch valves PV4, PV5, and PV6 are double acting. With system power off, both sides are open.

3. Air is supplied at rate of 30 bubbles/60s.

4. Syringe used to inject solid-phase reagent into reservoir is supplied with Technicon STAR System Reagent Kit.

5. Pump tube flow rates are applicable only for the Technicon STAR system. Flow rates for the same pump tubes when used on another system can be different.

[Reproduced with permission from Technicon Instruments Corp. (1980).]

have a 10-min incubation. The dwell time for STAR is relatively low—only 15 min.

Completion of the incubation time is detected by the "LAS error detector assembly," which monitors the analytical stream at the outlet of the incubator. The microprocessor activates the magnets, and the solid-phase antibody is trapped while the free antigen washes through. Pinch valve 6 is activated, directing the free antigen through a resistance coil to waste. A buffer solution flows into the analytical stream from above the first magnet and washes the bound antigen. Then the first magnet is switched off. The antibody then flows to the second magnet, where it is trapped and washed again.

When the second magnet is turned off, a vibrator is activated to dislodge the antibody particles. Pinch valve 6 is activated to direct the analytical stream through the coil in the scintillation detector. The computer accumulates counts for 45 sec.

The standard curve is fitted by a log-logit linear regression. The calculated concentration for each standard is compared to its assigned value; the standard curve may be rejected if the difference is excessive.

Operator–microprocessor communications are similar to those with SMAC. Performance has been described by Zborowski and Woo (1977) and by Karmel *et al.* (1980). Potential applications of STAR have been described by Forrest (1977).

4.   AUTOMATED HPLC USING SEGMENTED FLOW ANALYSIS SAMPLE PREPARATION

High-performance liquid chromatography (HPLC) has become an important technique for measuring concentrations of therapeutic drugs. The technique is somewhat labor-intensive owing to manual extractions and sample injection. Automated SFA extraction procedures have been available previously (Furman, 1976). In the fully automated sample treatment–liquid chromatography (FAST–LC) analyzer, (Technicon Instrument Corporation, Tarrytown, New York) serum specimens are sampled directly, extracted with solvent, and the evaporated and redissolved extract is injected into an HPLC system (Doland *et al.*, 1980). The sample pretreatment module of this analyzer is shown in Fig. 2. In the analytical cartridge, the samples are mixed with an internal standard and buffer and extracted into a mixed organic solvent. The organic solvent is too nonpolar for injection in the HPLC column, so the extract is evaporated. This is accomplished in the evaporation-to-dryness module. The extract flows onto the wire at point A, and the wire moves into the surrounding glass tube. A vacuum at point B draws heated air into the tube from point D, evaporating the extract to dryness on the wire. After leaving the glass tube, the wire passes through a pickup fitting at point E. Here, pickup solvent (methanol) washes the extract

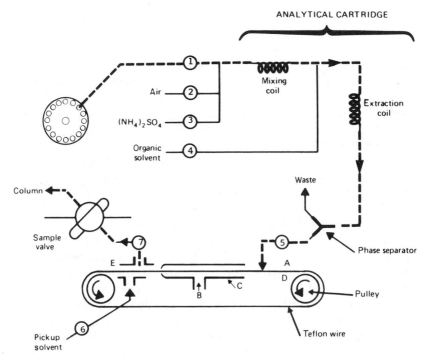

Fig. 2. Flow diagram of the sample pretreatment evaporation-to-dryness module for the FAST-LC system. [Reproduced by permission from Dolan *et al.* (1980).]

from the wire into the HPLC sampling valve. Pickup solvent is pumped away from the pickup fitting at a rate exceeding the rate at which it is pumped into the fitting; thus air bubbles are drawn into the stream to help maintain sample integrity. The bubbles are removed from the stream just prior to the sampling valve. Samples and injection valve phasing are controlled by the programmer. Internal standards are used for chromatographic timing and for troubleshooting, not for quantitation.

Analytical performance is comparable to manual HPLC and EMIT procedures according to Dolan *et al.* (1980).* The sample throughput is 7.5–10 samples per hour. Six anticonvulsants and metabolites were determined on each specimen.

B. *Flow-Injection Analysis*

When most clinical chemists think about continuous flow analysis, segmented flow analysis comes to mind. A new type of continuous flow analysis is just

---

*HPLC is also described in Chapter 5; FIA is described in Chapter 2.

beginning to be applied in clinical chemistry. Called flow-injection analysis, it differs from the conventional segmented flow analysis of Skeggs (1957) in several characteristics (Ranger, 1981). There is no segmentation by air bubbles in FIA, and the analytical stream is moving with laminar flow. Specimens are injected as a bollus by use of a sampling valve or loop rather than continuously entering the reagent stream. Thus, the system does not operate at steady state. Sample dispersion is controlled by minimizing hydraulic dead volumes. Very small bore tubing, low-volume detector flow cells, and low-volume tubing connectors are used.

FIA is developing into an alternative to traditional SFA. Sample dispersion is reduced enough to provide baseline resolution between samples at sampling rates of 90–120 per hour. Stabilization time is very short. FIA seems to be broadly applicable to different methodologies because all of the techniques of SFA can be adapted for use with FIA: dialysis, solvent extractions, immobilized reagents, and a variety of flow-through detectors. FIA is capable of precision and accuracy comparable to that of traditional clinical chemistry automated analyzers.

There are only a few applications of FIA to clinical analyses to date. One example is the system used by Renoe et al. (1980) to determine albumin with bromcresol green. The simplicity of the FIA procedure is demonstrated by the flow diagram shown in Fig. 3. The sample volume actually injected by switching the sampling valve is 2 μliter. The length of the timing coil was chosen to provide a 14-sec reaction time, in order to minimize interferences by globulins. A detector tracing taken at a sampling rate of 120 per hour is shown in Fig. 4. Approximately 10 sec are required for the detector signal to return to baseline following the peak maximum. Sampling rates of 180 per hour appear to be feasible with baseline resolution. The calibration curve relating the concentration to the integrated peak area shows a slight curvature.

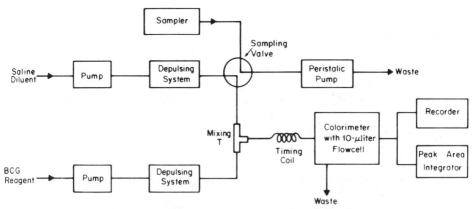

Fig. 3.   Flow diagram of FIA for the determination of albumin. [Reproduced by permission from Renoe et al. (1980).]

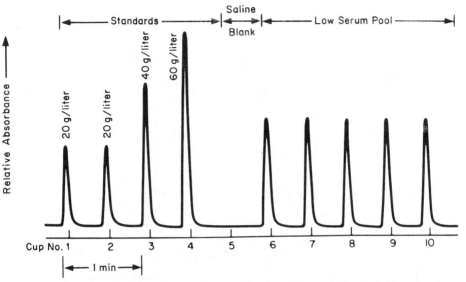

Fig. 4. Recorder tracing for analysis of serum albumin at 120 samples per hour. [Reproduced with permission from Renoe *et al.* (1980).]

An FIA procedure for total CO_2 has been reported by Baadenhuijsen and Seuren-Jacobs (1979). A segmented flow analysis method, including dialysis through a silicon–rubber membrane, was modified slightly. The sample volume actually injected was 50 μliter. Sampling was at the rate of 90 per hour.

High sample throughput with a clinically acceptable analytical performance can be achieved by FIA only if the sampling and valve timing are precise. Since a microprocessor is already necessary for measuring peak areas, it could also be used to coordinate timing for the sampling device. A microprocessor can perform many system-monitoring functions analagous to those of the Technicon SMAC computer described earlier. Kinetic enzyme measurements in unsegmented flowing streams have been demonstrated by Schlabach *et al.* (1979).

FIA has potential for other uses in clinical chemistry. Multichannel FIA systems appear feasible. FIA can serve as an automated precolumn extraction system for HPLC or a postcolumn reaction system.

III. Discrete Analysis in Open Tubes

A. *General Overview*

Automated discrete analyses are most commonly performed in systems using open tubes as the reaction vessel and in some instances as the cuvet. The design of many automated instruments is based on the process control of the mechaniza-

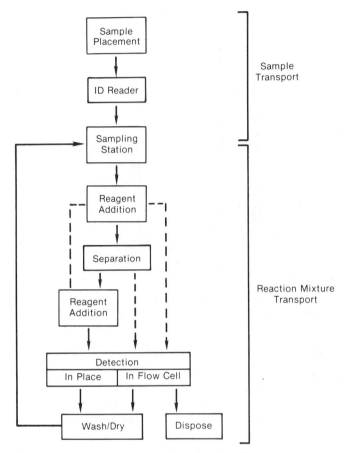

Fig. 5.   Flow diagram of automated analysis. →, transport function; ⋯→, optional transport.

tion and sequencing of manual manipulations. These operations are depicted schematically in Fig. 5.

A major advantage of discrete analysis using the open tube is that the tube may serve in a totally passive role. The tube does not dictate the type of method or reagents that must be used in the assay. The laboratory user has complete discretion in the selection of methods, within the confines of the hardware of course. On the other hand, specially designed tubes with immobilized enzymes or immobilized antibodies are available for specific assays, indicating the wide adaptability of the traditional test tube.

In general, automated discrete analyses in the open tube are not compatible with methods requiring separations, such as protein free filtrates. This aspect should not be ignored in light of the recent development of reference methods in clinical chemistry, many of which specify the use of protein free filtrates. In automated systems, however, the trend is toward greater specificity of reagents, obviating the need for sample pretreatment in routine testing.

There is a fairly even split among the discrete analyzers in the use of reusable tubes or disposable tubes. The higher volume systems use reusable tubes that are permanently mounted in the transport system in order to be readily available for the next sample. Often, they serve as the detection cuvet and must be of high optical quality with precisely matched path lengths. These systems, of course, rely upon a dependable rapid wash system to recycle the tubes. The mid-to-low-volume systems utilize disposable tubes and perform the analytical measurement in a separate detection cell.

There are basically two transport systems in any automated device, whether continuous flow or discrete (Fig. 5). One transport system sequentially presents the samples to the sampling device and the other transports the diluted sample–reagent mixture through the various stations to the detector.

A key component of the sample transport design is the sample identification system. The greatest flexibility in operation with minimized sampling errors is achieved with a positive ID reader, which reads the sample that is in position to be sampled. With this option, the requirements for carousel position identification are eliminated. Calibrators, controls, STATS, or routine samples are identified automatically by the ID reader. The ID reader enables the use of open-ended, linear sample transport systems as well. Without the ID reader, the process-controlled system must key on the carousel cup number. The sample transport system is programmable in many systems to facilitate the assay of calibrators, controls, and stats.

Transport jams and empty positions are detected with appropriate flags. Some devices will also switch the sampling system into standby until the problem is resolved or more samples are to be introduced.

Transport systems for the reaction mixture vary in complexity from the traditional carousel to the ''in line'' transport chain, to the parallel tube slat, which holds 30 cups in each slat (*vide infra*, Hycel M). The design and layout of the transport system are determined by the hardware of the instrument, such as reagent addition facilities, time delays, and detection techniques.

The digital pipet with a positive fluids displacement is used almost universally in discrete sampling systems. Details of its use and control vary significantly. Of prime importance is the variability of the sample volume to provide the user flexibility in design, modification, and selection of methods. This flexibility is achieved either by manual switch selection or by programming the sampler controlling microprocessor by means of the keyboard. If sampling is coupled with a positive ID system, the computer can control sample uptake according to the tests requested for that particular specimen. For some discrete analyzers, there is a flush out of the sample with appropriate diluent, while for others, the proper volume of sample is dispensed sequentially without diluent. Automatic detection of short samples and clots in the sampling process has not been widely available.

Digital processor controlled reagent dispensing units are based primarily on positive displacement techniques. However, another approach to reagent dispensing utilizes pinch valves or rotary valves that deliver reagents under pressure in proportion to the "open" time. Flexibility in reagent volume is achieved by direct switching or by keyboard commands to the microprocessor. Although many reagents show adequate stability at room temperature, some must be stored under refrigeration (4°C). The latter situation requires special consideration relative to the needs for precisely controlled reaction temperatures. The reaction chamber must facilitate rapid warmup followed by the maintenance of precise temperature control. It is the temperature of the reaction mixture itself, and not the bulk temperature of the heating bath, that must be monitored to assure proper reaction conditions for accurate assays (especially for enzymes, which display large temperature coefficients). This is an obvious particular problem for the "open tube" reaction vessel.

Readout measurements can be made while the reaction mixture is still in the tube on the transport system, or the reaction mixture can be aspirated into a flow-through cuvet. If colorimetric and fluorometric measurements are to be made directly while the reaction mixture is in a tube on the transport, the optical characteristics of the tube must be of high quality, and the path length and tube positioning must be precise. The transport system carries the reaction tube directly into the light path for the appropriate measurement. In other instances, the reaction mixture is aspirated into a special measuring cuvet for colorimetric or fluorometric detection, and into other devices for flame photometry and ion selective electrode potentiometric measurements. Special techniques in RIA utilize antibodies immobilized on the tube wall. After incubation of the sample antigen and tracer-labeled antigen, the tube contents are aspirated and rinsed and the radio-labeled antigen bound to the immobilized antibody is counted.

B. High-Volume Multichannel Analyzers

We have chosen two examples of 30-channel discrete analyzers to illustrate high-volume testing in open tubes. The Parallel (American Monitor Corporation, Indianapolis, Indiana 46268) has one of the largest throughput capacities available today. The instrument is capable of processing 240 samples per hour, selectively performing up to 30 tests per sample for a data output rate of 7200 per hour. The analyzer is designed around an array of 30 parallel analytical channels, each having its own sampling and reagent stations and spectrophotometer. Some of the basic components, such as the sampler and reagent dispenser, are similar to those found on the KDA, also manufactured by American Monitor.

Prior to sampling, the test request information must be entered into the computer along with at least the specimen tube number (or chain position number if an ID label is not used). Other demographic information can be entered into the computer as well. This can be done in either of two ways, mark sense cards or keyboard. The bar coded specimen tubes can be placed at random in the sample transport chain and are decoded by two bar code readers prior to the first sampling station. The computer then commands the appropriate sampling stations to take up the sample when the specimen arrives at each of the 30 sampling stations. The sample transport chain advances from one sampling station to the next every 15 sec. Thus, the sample for the thirtieth test in the system is taken up some 7½ min after the first test is sampled. Turnaround time can be less than 12 min.

The sampling system utilizes a Hamilton syringe (Hamilton Company, Reno, Nevada 89510) to aspirate the sample into a Teflon line. The sample and flushing water are then dispensed by activating a high-speed solenoid valve (5-μsec opening or closing action) connected through a T connection in the sampling line. Sample volumes can be varied from 1 to 100 μliter by changing the stop on the syringe and the open time of the high-speed dispensing solenoid valve. The sample and water diluent are dispensed into rewashable reaction tubes, which are transported away from the sample transport chain at right angles. The reagent dispensing units are programmed to dispense reagent only when tests have been requested. Additional reagents can be added at later times farther along the analytical channel as required by the method. The volume of reagent delivery is a function of the pressure of the pressurized reagent bottle, the tubing bore of the reagent line, and the time interval for which the high-speed solenoid valve is open. The nominal flow rate of reagents is 1 mliter/sec. The user may change the volume delivered by changing the open time of the valve in the computer.

The reaction temperature is maintained at 37°C with a water bath and is monitored in the water bath as well as in each spectrophotometer block by thermisters. The temperature of the spectrophotometer block is read and stored with the spectrophotometric data. Flags are activated if temperatures exceed acceptable limits.

The aspiration probe for the spectrophotometer can be placed along the reaction transport system according to the appropriate time for each particular test. The user informs the computer of the location of the spectrophotometer probe. Eight measurements are taken for kinetic methods at preset 1.2-sec intervals. The total measurement time is limited by a maximum residence time of 12 sec. The data processing programs examine the data for linearity and may exclude one of the eight data points based on a defined multiple of the standard error of the regression line of the absorbance data. If more than one point exceeds acceptable limits, the computer alerts the operator of the analytical error, and the assay should be repeated. End-point methods use ''quadrametric'' photodetection with four interference filters to generate a predictive curve of the sample blank for more accurate blank compensation. Measurements for sodium and potassium are obtained with a flame photometer. The sample is diluted with a lithium internal standard solution and is aspirated into the flame photometer. After measurement, the reaction tubes are emptied, washed, and dried as they are recycled beneath the analytical system for reuse.

The recommended frequency of calibration is once per day. A method may use from 2 to 12 calibrators to define the calibration line. The various curve fitting programs available include point-to-point, linear regression, and various curvilinear and polynomial functions which find application in enzyme immunoassays.

There are three levels for accepting patient data into the files. For transition acceptance, all patient sample data must be bracketed by quality control specimens whose results are within acceptable ranges. Otherwise, the patient data are held for the operator to review in the manual acceptance mode, to be repeated if necessary. Transition acceptance also puts a temporary hold on data that are above or below linearity limits.

Manual acceptance enables the operator to review data that are being held in temporary files for the above-mentioned causes. The operator would be overriding these flags to file data from the manual acceptance mode.

Automatic acceptance is used to file data that have been reviewed off-line for errors that have been corrected. Automatic acceptance is also used when external data from other analytical systems in the laboratory are being entered into the computer; in this case the computer is being used as a central laboratory computer to generate comprehensive laboratory reports. The computer can store data for up to 48 different tests per sample. This collating function is relatively common in analyzers with microprocessors.

There are a number of utility programs available that enable the user to adapt the system to his needs. These include the definition of normal ranges (by age or by sex), panic values, profiles, the definition of up to 16 calibrators (their values and expected absorbance ranges), and expected ranges for up to 16 quality control materials. Other computer features include the display of system parameters, such as incubation temperature, high or low absorbance values, panic results, transport status, and the status of complete/incomplete samples. In the event that repeat analyses are necessary, the tube is merely replaced on the sample transport for resampling and the operator may then file the new corrected result for that single test that was incomplete. If dilutions are necessary, they are performed by placing an uncoded specimen tube containing the diluted specimen into the sample transport chain. The operator keys in the particular chain number with the specimen number, tests, and dilution factor. The computer automatically performs the required tests and calculates the final results. The data files can be reviewed by means of specimen number (bar code), by patient name, by physician name, or by date. As of this writing, we have found no reports of performance data for the Parallel in the literature.

The progress in process-controlled discrete analyzers over the last decade is demonstrated in the series of automated analyzers marked by Hycel Inc., Boehringer Mannheim Company, Houston, Texas 77063. The series begins with the 10-channel Mark X and then the 17-channel Hycel 17 and Super 17. The latest model in the series is the 30-channel (25 tests and 5 blanks) Hycel M.

The Hycel M utilizes two microprocessors. One microprocessor is used for controlling and monitoring the system functions (MOM, or machine oriented microprocessor), and a second microprocessor is used for data processing and patient files (POP, or patient oriented processor). Operator communication with POP is facilitated by means of a coded keyboard, a CRT display, an integrated mark sense card reader/printer, and a separate 200 character/sec lineprinter. A floppy disk storage system is capable of storing results from 1500 samples per disk.

Prior to sampling, the test request information and specimen identification must be read into the computer by a card reader for test programming. The samples can then be placed randomly in the sample transport chain, which advances every 30 sec. The sample tube bar code is read just prior to sampling. STAT samples can be placed in the transport chain just before the bar code reader. Analysis turnaround time is 15 min. It is possible to key the specimen identification number into the computer by means of the keyboard in the absence of bar codes.

Sufficient sample is aspirated by the sample for all the programmed or selected tests, up to a maximum of 2 mliter. The programmable, positive displacement pipetting system then sequentially dispenses the appropriate volumes of sample (7.5–90 μliter) into the respective tubes mounted in a 30-tube slat for parallel transport to subsequent stations. The reaction tubes advance to a 37°C water bath. Diluent and reagent are added by programmable positive displacement pumps at various selectable positions during the next 6 min or 12 positions. After 10 min, 0.6 mliter of each reaction mixture is aspirated into separate photometric flow cells (or flame photometer for sodium and potassium). The optics for each channel consist of a beam splitter placed behind the illuminated sample cell to direct the transmitted light through two separate interference filters (for bichromatic measurements) onto a reference and a test photodiode detector. Thirty-two measurements are taken for each method at fixed half-second intervals. The slope for the absorbance versus time curve for kinetic methods is determined by regression analysis with limit checks for outliers. Data in nonlinear regions are identified by use of second derivatives and rejected. After measurement, the mixture in the measuring cell is rinsed back into the reaction tube. After the tube advances out of the 37°C bath, it is inverted to dump the mixture, rinsed, washed, dried, and cooled—ready for reuse at the sample dispensing station.

Calibration and calibration update uses a two-point approach and can be done at user-defined frequencies or randomly by placing the calibrator on the sample transport chain. The calibrator has a special bar code, which instructs the computer accordingly. All or selected tests can be calibrated as determined by the user, Out-of-range blank and calibrator data are flagged to alert the operator for corrective action.

The system monitor displays the status of the transport systems and the temperature of the incubation bath and spectrophotometer section. Expected ranges for up to six quality control materials can be stored in memory for on-line quality control monitoring. Out of range results are flagged. Quality control data are stored in the files and are available for summary calculations. Patient data can be printed out on the small mark sense card reader/card printer in random order or on the medium-speed line printer as the sample analysis is completed. The medium-speed line printer is also used to print quality control data and other reports, generated in the computer. The Hycel M can be interfaced to a central laboratory computer. At present, we have not found a formal evaluation of the performance of the Hycel M reported in the literature.

C. Single-Channel, Sequential, Multitest Analyzer

A new variable test discrete analyzer has recently been developed from the initial design concepts of Snook *et al.* (1979) and Mitchell (1980). It is called the discrete analyzer with continuous optical scanning (DACOS, Coulter Electronics, Inc., Hialeah, Florida 33014). The instrument uses eight mobile colorimetric detectors rotating one revolution every 6 sec around a 120 cuvet carousel, which itself indexes one position every 6 sec. There is a tungsten halide source at the center of the carousel. Each optical channel emanating from the source lamp consists of a light pipe, interference filter, and detector. This design allows rapid indexing of tests and increases the flexibility in analytical procedures. Serum blank absorbances can be obtained prior to the addition of reagent. A second reagent can be added 2 min after the first one. Since readings are obtained at each of 100 positions (20 positions are used by the wash and dry station), the analyzer is suitable for the very fast endpoint methods as well as the slow kinetic methods that may require up to 10 min to monitor the change in absorbance. Superimposed on this is the ability to measure the absorbance as a function of up to 8 wavelengths at each index position—to monitor side reactions or to obtain other blanking information.

The mechanics of the instrument consists of three carousel units (Fig. 6): the 120 cuvet reaction carousel already mentioned, a 4 concentric ring carousel for samples (64), STATS (16), and calibrators and controls (16), and a double ring carousel for reagents (12 reagent containers per ring). The sample carousel is indexed only after aliquots for all tests for a sample have been dispensed sequentially into the reaction carousel. The reagent carousel rotates under computer control to the appropriate reagent for that test in the reagent dispensing position.

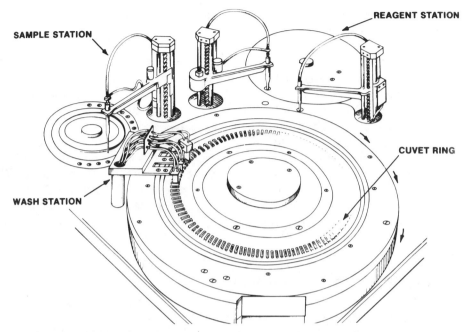

SAMPLE STATION

REAGENT STATION

CUVET RING

WASH STATION

Fig. 6. Schematic view of DACOS. The rotating light source is under the center cover. Around it rotate interference filters for eight different wavelengths, one revolution every 6 sec. As the cuvet ring indexes one position every 6 sec, cuvets pass by the sample station, reagent stations and finally to the wash station. (© 1980 Coulter Electronics, Inc., reproduced with permission.)

The sampler is electromechanically operated and can sample volumes from 2 to 20 μliter, variable in 1-μliter steps. The reagent dispensers provide volumes of 80 to 500 or 20 to 200 μliter. The reagent dispensing system does not use a flow-through system but rather is based on the positive displacement "sipper" principle to avoid pump priming for each reagent. After dispensing the reagent, the probe vibrates in the reaction mixture to mix the sample and reagent together. The reagent probe is then rinsed off in a wash receptacle prior to picking up the next reagent for the next test. This eliminates sample carryover and reagent contamination, since one probe is used to dispense all reagents at each reagent station.

Of the 100 readings per test (if single wavelength), only those pertinent to the type of analysis as defined by the user are stored in memory and are used to calculate analyte concentration or activity.

Calibration can be performed with as many as six calibrating materials. Linear regression is used to determine the calibration line for the linear methods using multipoint calibration. Other math routines are available in memory for the nonlinear immunochemical methods. Enzyme methods use a conversion factor based on the molar absorptivity of product or reactant to convert the change in absorbance to units to enzyme activity. The frequency of recalibration for specific tests is defined by the user and may involve reassay of all the calibrators or just a reagent blank to monitor and correct for baseline drift.

The user can define acceptable ranges for the quality control materials. Out-of-range situations are flagged on line. The quality control data can be stored as individual data to provide quality control charts for up to 60 days. Summary statistics can also be calculated for each material.

These functions are controlled by a minicomputer. The testing sequence is programed by the user with STAT interrupt capabilities. No published evaluation data are available at this time. However, with the relatively high rate of sampling and 10-min analysis time, it would appear to be compatible with both the emergency and intermediate volume routine testing needs of the laboratory. With the polychromatic analytical capabilities, the potential is also there to determine several analytes in one cuvet.

D. Batchable, Single-Channel Analyzer

Most of the discussion thus far has been about large automated analyzers. These large units are seldom seen in the smaller hospital laboratories, where the volume of testing may not justify them economically. These smaller laboratories need to automate their testing also, but on a smaller scale. Any automation they acquire must be very flexible, because there often is only one automated instrument in the laboratory. All clinical chemical tests to be automated must be automated on that one analyzer, which is usually a single-channel system. That analyzer must be economically efficient for runs with small batches of samples. It must be easily and rapidly changed from method to method. The analyzer must be able to accommodate a wide variety of methodologies. The applications of this broad group of analyzers are not limited to smaller laboratories. Most large laboratories have multichannel automated analyzers but still need the smaller, flexible analyzers for tests not included on their multichannel analyzers, or for tests requested singly, or for low-volume tests.

Some laboratories use a centrifugal analyzer to meet this need. Others use one of the variety of instruments ranging from spectrophotometer add-ons to completely microprocessor controlled systems. One very popular analyzer in this group in the past has been the ABA-100 (Abbott Diagnostics Division, Dallas, Texas 75247). Two successors to the ABA-100 are the Abbott VP and the ABA-200, which has been evaluated by Elser and Garver (1981).

The Abbott VP is a discrete, single channel, batch analyzer that can fit onto a laboratory benchtop. There are two modules in the VP, the control module and the processor module. The control module contains the minicomputer, electronic

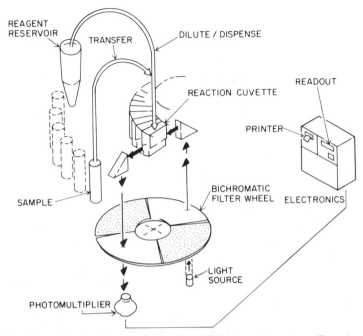

Fig. 7. System function diagram of the Abbott VP Bichromatic analyzer. [Reproduced with permission from Abbott Laboratories (1980).]

circuitry, and the printer. The processor module is the site of sampling, reagent addition, incubation, and readout measurement. Functions of the processor module are shown in Fig. 7. Sample cups are placed around the periphery of a round carousel, which rotates on the photometer housing. Reaction cuvets corresponding to each of the 32 sample cups are contained in a circular plastic multicuvet, which is the middle ring of the carousel. It is thermostatted by a water bath. The center of the carousel is the chimney from the light source below. The carousel spends either 3.75 or 5.625 sec at each sample position. Sampling and primary reagent addition are accomplished by means of a single probe operating from a gear head assembly that positions the probe in the sample cup. The sample is drawn into the probe and the reagent is drawn into the reagent syringe by stepping motors. The probe is positioned in the corresponding cell of the multicuvet, and the sample is washed into the cuvet by the reagent. A second reagent may be added by using an auxiliary probe on the first or later rotation of the carousel, as programmed by the user. The current position of the carousel is given by an optical binary code in the carousel skirt. Incubation takes place in the cuvet. The carousel makes one complete revolution every 2 or 3 min, as programmed.

Colorimetric readout takes place in a fixed photometer unit that reads each cuvet when it is in the light path (once per carousel revolution). The ABA-100, VP, and ABA-200 utilize bichromatic photometry to improve measurement accuracy by self blanking. A minimum of two carousel rotations are required for end-point chemistries, and three rotations for rate methods. Sample throughputs of 465 per hour are theoretically possible. The Abbott Executive accessory enables the collation of test results and printing of patients' reports. Performance data for the VP have been reported by Fu et al. (1979).

E. Radioimmunoassay in Open Tubes

Since radioimmunoassay (RIA) testing is done manually in open tubes, it was natural to mechanize RIA by operations closely resembling the manual manipulations. The basic steps in an RIA procedure are incubation of standard or patient sample with antibody and radio-labeled antigen, separation of antibody bound antigen from free antigen, and counting of the radioactivity of either the free or bound fraction. The first fully automated system for RIA was the Concept 4 (Micromedic Systems, Horsham, Pennsylvania). This instrument has been described by Johnson et al. (1976) and Chen (1980a). It uses antibody-coated tubes to affect the separation of bound and free antigen. The tubes move through the analyzer in racks. Specimens are pipetted into the coated tubes, radioactive antigen is added, and the tubes are mixed. The rack of tubes moves into the incubator and remains for a programmed time (up to 17 h). The racks advance to the aspirate-and-wash station, where free antigen is removed from the tube. The tubes are then transported to the scintillation counter. Data reduction is on-line with a programmable calculator. Up to 200 tubes can be processed per run at throughputs of up to 240 tubes per hour. Another run of the same or a different test can be started as soon as the previous run leaves the incubator. There is provision for setting up the new run before the previous run is complete. The system is modular, and different sections may be used individually when one section is not functioning. Performance data have been reported by Painter and Hasler (1976), Carter and Grimmet (1976, 1977), Button and Gambino (1977), Lawn et al. (1977), and Chen (1980a,b).

Many commercially produced manual RIA kits achieve separation of the bound and free antigen by precipitating the antibody from solution with reagents, such as the second antibody, polyethylene glycol, ammonium sulfate, or some combination of these. Sometimes, charcoal or talc is used to adsorb (and thus separate out) the free antigen. Recently, kits using antibodies chemically bound to glass beads or polymers have become popular. In nearly all these procedures, specimen and radio-labeled antigen are incubated with the antibody reagent in test tubes. Then the separating agent is added and incubated for a short time. The bound antigen is then separated physically by centrifugation; the precipitated phase forms a packed button on the bottom of the tube. The supernatant is separated, and either the precipitate or the supernatant is counted.

Bagshawe (1978) developed an automated system that enables direct automation of many of these manual RIA procedures. This system is the basis of the PACE 4 (Picker Corporation, Northford, Connecticut 06472), described by Bowie (1980) and Chen (1980a). It provides flexibility for the user to choose reagent kits and adapt them to the analyzer. In operation, samples are aspirated and transferred into reaction tubes in a reaction tray. The tray is then moved to a refrigerated reagent dispensor, with nine syringes for different reagents. The contents of the syringes are constantly being mixed. After reagent addition, the tray moves to an incubation chamber large enough to permit runs of up to 480 tubes. Incubation may be set at 37°C, 24°C, or 4°C. After incubation, precipitants or solid phase adsorbents may be added, if necessary, by transporting the reaction tray back to the reagent dispensing station. Then the tray is moved to the separation module, where the contents of each tube are aspirated and filtered through a glass fiber filter pad mounted on a continuous tape of 1000 pads. The pad now contains the precipitate or charcoal. The pad is moved by the tape through four counters in sequence, spending 15 sec in each. The sum of all four counts for each filter pad is used in calculations. Thus, 1-min counts are obtained, at the rate of 240 samples per hour. The microprocessor performs calculations of patient results by the user's choice of curve-fitting programs.

Adaptation of RIA procedures to the PACE 4 have been reported by Piknosh *et al.* (1979) and Dass *et al.* (1980).

Open test tubes and a flow-through separation system are combined for automated radioimmunoassay in the ARIA II (Becton-Dickinson Immunodiagnostics, Orangeburg, New York). It operates in three modes: simultaneous, sequential, and incubation (Bowie, 1980; Chen, 1980a). In the simultaneous mode, sample and radioactively labeled antigen are aspirated, mixed, and pumped into a sampling loop. The contents of the sampling loop are then washed with incubation buffer into a chamber containing antibody covalently bound to glass beads or fibers (Chen, 1980c). The chamber is flushed with buffer, carrying the free antigen to the detector for counting. Then an elution buffer washes the bound antigen from the chamber and into the detector for counting, and at the same time, the sampling loop and sample aspirator are backflushed to reduce intersample mixing. Then the antibody chamber is rinsed with incubation buffer to prepare for the next sample.

In the sequential mode, the radio-labeled antigen is injected into the antibody chamber after the sample has been injected. Sequential addition of the sample and radio-labeled antigen increases sensitivity.

In the incubation mode, the sample, radio-labeled antigen, and antibody are preincubated before being introduced into the flow system. There are two rings of plain cups on the carousel. Cups in the inner ring contain samples, and corresponding cups of the outer ring are the site of preincubation. The specimen, radio-labeled antigen, and antibody are dispensed into an outer cup for each sample. After a specified incubation period, the mixtures in the outer cup are analyzed in the simultaneous mode, using the chamber to separate bound and free antigen. Maximum throughput is 30 samples per hour. Incubation times are variable. Performance has been evaluated by Reese and Johnson (1978) and Chen (1980b,c).

IV. Discrete Analysis by Centrifugal Analyzers

The centrifugal analyzer, developed by Anderson (1969) at Oak Ridge National Laboratory, has been called "the third generation of automatic analyzers" (Mitchell, 1980). Segmented continuous flow systems are considered the first generation and automated discrete tube systems, the second generation. The unique capability of the centrifugal analyzer is the multipoint detection of reagent blanks, standards, and samples on successive revolutions. This parallel mode of operation eliminates the effects of instrumental drift in the analysis.

The five manufacturers of centrifugal analyzers are listed in Table II. The principles of centrifugal analyzers and their performance characteristics have been reviewed in a special publication by the American Association for Clinical Chemistry (Savory and Cross, 1978). The centifugal analyzer consists of a sample–reagent dispensing system, an optical cuvet assembly mounted on a centrifugal rotor, and a computer module that controls the centrifugation and data acquisition processes, performs the appropriate calculations, and prints out the results. The sample–reagent dispensing systems vary from stand-alone electromechanical units to microprocessor-controlled systems integrated with the rotor module. Rotors are designed from two basic approaches: the rewashable combination of Teflon™ transfer disk and quartz optical head or the disposable acrylic plastic rotor that serves as the reaction vessel and cuvet. Information about the rotors of the currently available systems is in Table II.

All systems perform the typical colorimetric and turbidimetric determinations. Additionally, the IL Multistat III FS/LS is also capable of fluorescence (Pearson et al., 1981) and nephelometric measurements (Hills and Tiffany, 1980). The Cobas-Bio system uses a uniquely designed horizontal cuvet. This feature eliminates the dependence of the analytical signal on diluent or reagent volume (Eisenweiner and Keller, 1979). Although this should result in an inherent improvement in precision, the analytical performance has been reported to be quite similar to that of other centrifugal analyzers (Parker and Cross, 1980; de Graeve et al., 1981).

Centrifugal analyzers are generally operated in the traditional batch mode,

TABLE II

CENTRIFUGAL ANALYZERS

Name	Manufacturer	Rotor capacity	Rotor type[a]	Detection types[b]
Centrichem 600	Baker Instruments Corp. Pleasantville, New York	30	R	A, T
Cobas Bio	Hoffman La Roche, Inc. Nutley, New Jersey	30	D, horizontal	A, T
Gemsaec	Electro-Nucleonics Inc. Fairfield, New Jersey	16	R	A, T
GEMINI	Electro-Nucleonics Inc. Fairfield, New Jersey	20	D	A, T
FLEXIGEM	Electro-Nucleonics Inc. Fairfield, New Jersey	20	D	A, T
Multistat III	Instrumentation Laboratory, Inc. Lexington, Massachusetts	20	D	A, T
Multistat III FS/LS	Instrumentation Laboratory, Inc. Lexington, Massachusetts	20	D	A, T, N, F
Rotochem II	Travenol Laboratories, Inc. Savage, Maryland	15	R	A, T
Rotochem IIa	Travenol Laboratories, Inc. Savage, Maryland	36	R	A, T
Rotochem CFA 2000	Travenol Laboratories, Inc. Savage, Maryland	36	R	A, T

[a] R is a reusable, Teflon transfer disc and quartz cuvet (Teflon is a trade mark of DuPont Co., Wilmington, Delaware). D is a disposable, UV-transmitting acrylic cuvet disc.

[b] A is absorbance, T is turbidimetric, F is fluorescence, and N is nephelometric.

each batch for a particular analyte. Travenol Laboratories (Savage, Maryland 20863) has recently developed software to enable the determination of as many as six different analytes in a single run. This facilitates the analysis of STATS and converts the centrifugal analyzer from a strict batch analyzer to a limited variable test analyzer. The menu for each analyte includes reagent blank, standard, controls, and unknowns as appropriate. The Cobas-Bio system also offers unique flexibility in handling small batches (such as for STATS). Its microprocessor monitors the use of cuvet spaces on the 30 space rotor. The unused cuvets are available for subsequent batches, which can be loaded by the integrated sampling unit while the cuvet rotor remains in place. A feature also unique to the Cobas-Bio (but which could be incorporated into any system using disposable cuvets) is the ability to take an initial absorbance reading, stop the rotor, add a second reagent, and then take final readings.

Major differences in the centrifugal analyzers appear in the variety of calibration algorithms that are available in the system software. Every system offers the basic endpoint programs based on a single calibrator with or without a zero setting and with or without a reagent blank. Certain systems offer the capability to assay several calibrators from which an average conversion factor (i.e., slope of the calibration curve) is calculated. This approach assumes a zero intercept in the method, which may or may not be the case. The ideal calibration software for endpoint methods should provide enough flexibility for the user to select a variable number of calibrators up to, say, six. Sometimes no calibrator is available, and a theoretically derived conversion factor must be used. Suitable calculations should include a linear regression estimate of the calibration line for the multipoint calibration procedures (for example, Garber and Miller, 1980).

In the kinetic mode, there is also a basic variety of measurement and calculation programs, which include the two point measurements and multipoint measurements with user selectable read times and intervals. The number of multiple readings is limited by some software to 9 readings and in others up to 30 readings.

Application of centrifugal analyzers in the clinical laboratory has grown rapidly over the past decade. Their natural application is for kinetic assays. However, data acquisition and data processing software have been developed for the more sophisticated curve fitting algorithms applicable to enzyme immunoassays and fluorescent immunoassays (Pearson *et al.*, 1981).

V. Discrete Analysis with Prepackaged Single-Test Reagents

All of the automated instruments described thus far are designed to operate optimally in an environment in which many specimens arrive over a relatively short time period to be analyzed in a batch. Most automated analyzers are not efficient when performing a variety of different tests on the few samples that come into the small laboratory as routine or STAT tests or into the larger laboratory on a STAT basis. Reagents are usually made up daily for each test in quantities for many specimens. Often, there is a startup sequence during which reagents must be pumped through the analyzer for a while before instrument operation is stable. Standards and controls must be run with each batch of patient specimens (or with each specimen tested individually). Thus, running one emergency specimen by itself can require the same complete setup and standardization as a batch of specimens, and almost as much reagent. Each additional different test requires its own additional reagent, an additional analytical channel, and possibly additional setup time. The same sorts of inefficiencies apply to performing the tests manually on an emergency specimen. In addition, operator training becomes more of a limiting factor for manual performance of individual emergency testing, especially with the increasing variety and complexity of testing required on a STAT basis.

One successful solution to the problem of analyzing single specimens is to have reagents for performing a single analysis on a single specimen premeasured and packaged as separate units. The packaging also acts as the ''cuvet'' for detection of reagent; its package should be stable indefinitely, and there should be no variation from package to package within a manufacturing lot of reagent. Thus analyzer calibration should be stable over the duration of the entire manufacturing lot. Then the analyzer itself could be as easy to operate as an office copier—the operator loads the specimen(s) and tells the analyzer which tests to perform. Tests would be run in any sequence. This approach has been developed in the automatic clinical analyzer (aca, E.I. Du Pont de Nemours & Company, Wilmington, Delaware 19898) and the Ektachem 400 (Eastman Kodak Company, Rochester, New York 14650).

A. Single-Test Analysis Using Disposable Reagent Packs

The Du Pont Company has produced a series of discrete automated analyzers, beginning with the aca I in 1970. Subsequent modifications led to the aca II in 1976, which involved changes in the dedicated computer to improve operator interaction with the computer. In 1978, the aca II was expanded from a capacity of 30 tests to 60 tests. Also in 1978, the microprocessor-controlled aca III was introduced. Its analytical performance was studied by Garber et al. (1979) using evaluation protocols proposed by the National Committee for Clinical Laboratory Standards (NCCLS) (1979).

The operational principle of all aca's involves the prepackaged reagent pack. Each method is identified with a human readable code and an instrument readable 6 position binary code ($2^6 = 64$) to give a capacity of 64 different tests. The appropriate test packs are selected by the operator and placed in a sampling rack behind a sample cup labeled with patient identification. The instrument advances the sample cup to the sampling position, decodes the first pack in order to identify the buffer type and sample volume, and dispenses them into the pack. The pack is advanced into a 37°C (or 30°C) air bath and onto the transport chain. Every 37 sec (at 37°C), or 54 sec (at 30°C), the transport chain advances one position. In the first two positions, the pack passes through a preheater to bring the pack to 37°C (or 30°C). A mixing station (Breaker/mixer) is located at position 3. This breaker/mixer presses and squeezes the pack to mix the diluted sample with reagents. The reagents are heat sealed in seven small compartments; the first four are opened at the first breaker/mixer. Positions 4–8 are incubation stations. Position 9 is the location of the second breaker/mixer; the last three reagent compartments are opened there. Thus, a second group of reagents can be added 6 stations later or about 3¾ min later. The photometer is located at position 10. Again the pack is decoded and verified by comparing it with the code determined at the filling station. The measurement can be either endpoint or kinetic. The endpoint methods utilize two wavelengths or two packs (one blank, one test). The kinetic measurement takes two readings 17 sec apart. Conversion of absorbance to concentration or activity units is computed according to the stored intercept and slope factors for linear methods or by means of power functions for the nonlinear immunoassay methods. The pack is discarded from the transport chain at position 11. The readout also appears at this time. The total turn-around time is less than 7 min for most methods at 37°C.

To implement microprocessor control on the aca III, all driving motors were replaced with stepping motors. An 8×32 character display was included in the aca III to provide status displays of all moving and analytical components, to give visual readout of results, and to display operator input and instructions to memory on the alpha numeric keyboard. A floppy disk stores the instrument paramenters for up to four versions of each of 64 different methods. The particular version that is in active use is stored in the 32 K RAM memory.

The basic operation of the aca III is similar to its precursors, however the instrument has been supplemented with a number of function sensors to assist in the identification and troubleshooting of system malfunctions, whether they are in the sampling system, the reaction chamber components and transport system, the photometer, or the printout device. These system sensors also find use as part of maintenance prompting routines. They can prompt the operator with the next step in a particular maintenance procedure. Examples are the precise adjustment of the sampling needle or the zero adjustment on the digital sample–reagent pump. Other maintenance procedures controlled and monitored by the microprocessor include the priming of the various reagent lines, and the cycling of the filling station, the two breaker/mixers, the photometer, the transport system, and printer to demonstrate correct function.

The microprocessor assists in the recalibration of methods by performing the calculations and automatically filing the new factors in memory as desired by the operator.

The user is dependent upon the manufacturer for the development of new tests and for consistent quality of reagents. Analytical performance data have been provided in many evaluations (for example, Westgard and Lahmeyer, 1972) and in external proficiency surveys (Ross *et al.* 1980).

A recent addition to the aca system is an ion-selective electrode accessory, which is interfaced directly to the microprocessor. Thus, no special steps are required by the operator to use these tests. At present, sodium and potassium are available on the unit, which has a capacity for six electrode methods. The analytical performance of the ion selective electrode system has been reported by Miller *et al.* (1980).

B. Single-Test Analysis Using Thin-Film Reagent Slides

Reagents for a single test are contained in a thin, dry film package analagous to a photographic slide in the Ektachem analyzer developed by Eastman Kodak. These slides were described in general by Curme *et al.* (1978). A cross section of a colorimetric slide is shown schematically in Fig. 8. The film consists of several layers. The specimen is aspirated from a sample cup and deposited as a droplet onto the top spreading layer to be transferred uniformly into the reagent layer. The reagent layer contains analyte-specific reagents bound by hydrophobic polymers. As the sample is incubated at 37°C, color develops in proportion to the analyte concentration in the specimen. It is interesting that uniform delivery of the fluid into the reagent layer by the spreading layer causes the area at the center of the slide to have a concentration that reflects the specimen concentration faithfully in spite of variations in sample volume. A change of 10% in the sample volume causes only a 1% change in the area concentration within the center of the reagent layer. The color is measured by reflectance densitometery. A nonlinear calibration curve of reflectance density versus concentration results; it is linearized by the use of the Williams-Clapper (1953) transform. The composition and analytical performance of the glucose slide was described in detail (Curme *et al.*, 1978).

Application of this technology to several clinical chemical analyses was de-

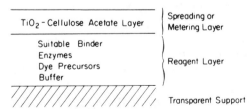

Fig. 8. Schematic cross section of an Ektachem analytical film. [Reproduced with permission from Curme *et al.* (1978).]

scribed by Spayd *et al.* (1978). Methods for urea, amylase, bilirubin, and triglyc-
erides were reported. Composition of the spreading and reagent layers were
changed to produce the desired analyses. The application of multiple reagent
layers and a semipermeable membrane to the slide analysis system was described. Analytical performance data accumulated according to the protocol of
the National Committee for Clinical Laboratory Standards (NCCLS) (1976) were
demonstrated. Evaluations of the Kodak Ektachem GLU/BUN Analyzer were
reported by Warren *et al.* (1980), Cate *et al.* (1980), and Bandi *et al.* (1981a,b).
Simultaneous determination of total and direct bilirubin on a single slide was
reported by Wu *et al.* (1980).

For measuring electrolyte concentrations, potentiometric methods have been
adapted to the slides. As with the colorimetric (reflectance) methods, each slide
is used for one test and then discarded. The potentiometric slides have been
described in general by Curme *et al.* (1979). Specific descriptions of the sodium
and total CO_2 slides have been given by Daniel *et al.* (1980) and Kim *et al.*
(1980), respectively. Each potentiometric slide consists of a pair of electrodes.
The electrodes are thin films of silver–silver chloride on a polyester base. Each
silver–silver chloride film is covered by a hydrophilic polymer matrix containing
salts. This is then covered by an ion-selective membrane. A concentration cell is
produced on each slide by two nearly identical electrodes connected by a porous
paper bridge. In use, a potential is developed across the cell by placing a drop of
reference solution on one electrode and a drop of the specimen on the other. The
fluids flow into liquid contact by means of the paper bridge. Drops are dispensed
and touched off onto the slide by the same pipetting system as that used for
colorimetric slides. The slides are stored dry and reach a stable potential within a
few minutes after the fluids are placed onto the dry electrodes. Nernstian response is obtained. Performance data from an evaluation of manual and automated analyzers utilizing this slide technology for electrolyte measurements have
been reported by Schnipelsky *et al.* (1979). The results of an evaluation of an
electrolyte analyzer according to NCCLS protocols (NCCLS, 1979) were reported by Powers *et al.* (1980). Comparison data for potassium and chloride
were also reported by Simkowski *et al.* (1980).

The most recently developed instrument that uses the Kodak slide technology
is the Ektachem 400, described by Blake *et al.* (1980). Twelve tests were
developed at that time. The analyzer operates at 300–540 tests per hour. Colorimetric tests require approximately a 6-min turnaround time, and potentiometric
tests require approximately 4 min. The entire unit is under microcomputer control. The operator may order panels or individual tests at the control unit.

The traditional round sampler trays are split into four crescent-shaped sections. Each section can be loaded and run individually, permitting continuous
sampling, because tray sections can be removed after sampling and replaced with
new sections without interrupting the sampling process. The entire tray of 40

samples can be handled as a unit, if desired. Samples are loaded onto the tray in disposable cups, each accompanied by a disposable pipet tip for metering the sample onto slides. The pipet tips are mounted in holes next to the sample cups. Any sample to be tested STAT is accessed by the sampler as soon as it has finished metering the present sample.

Slides are stored in cartridges from which they are dispensed as needed at the time of use. The top of each cartridge is labeled with human and machine readable codes to identify the chemistry. Cartridges are stored in the analyzer in one of two slide storage modules, each of which contains up to 16 cartridges. Maintained below 23°C, they are stable for at least 1 week. The cartridge labels are optically scanned by the computer as the storage module rotates until the correct cartridge is in place to dispense the desired chemistry slide into the metering station. Thus, slide cartridges may be loaded randomly.

During the metering cycle, the pipet tip previously placed on the sample tray is picked up by a tapered mandrel. The sample is aspirated, and a 10-μliter drop is dispensed by air displacement and touched off onto the spreading layer of the slide. For potentiometric tests, a drop of reference solution is dispensed onto the reference part of the slide by a similar metering system. When all tests requested on a particular specimen have been started, the pipet tip is ejected and the analyzer is prepared for the next specimen.

The slides are processed into an incubator. They remain there for 5 min at 37°C for colorimetric tests or for 3 min at 25°C for potentiometric tests. Slides pass through the reading station several times during the incubation, to facilitate shorter incubation times and multiple readings of the same slide. Recalibration is performed weekly.

Performance data from an evaluation of six colorimetric tests and four potentiometric tests on the Ektachem 400 according to NCCLS protocols EP2-P, EP3-P, and EP4-P were reported by Rand *et al.* (1980).

VI.   Discrete Analysis by in Situ Techniques

The classification of techniques based on the in situ analytical approach includes a very broad range of analyzers varying widely in the degree of automation. In situ analyses refer to those in which a sample or sample–reagent mixture is aspirated or dispensed directly into the measuring cell. Additional preprocessing steps by the analyzer such as separation, incubation, and the necessary transport between such stages are eliminated. Thus, the total turnaround time is limited solely by the analysis time. This approach is particularly conducive to the rapid turnaround needs of the clinical laboratory for emergency or STAT analyses. However, it places major limitations on the maximum allowable time for the reaction and measurement of an analyte. The slower kinetic methods for determining enzyme activity have not found widespread application with this technique. Both photometric and electrochemical methods have been used for detection.

A. Modular Multichannel in Situ Analyzer

The ASTRA 4 and ASTRA 8 (Automated STAT/Routine Analyzer, Beckman Instruments, Inc., Fullerton, California 92634) are discrete programmable analyzers with either four or eight separate measuring cuvets. The instruments are modular in design. The user may select four or eight chemistry modules from among the 12 different methods presently available. Although the 4- and 8-channel analyzers are similar in most respects, the ASTRA 8 does have a CRT display as standard equipment and an accessory floppy disk for data storage, which are not available with the ASTRA 4. Communication with the microprocessor is achieved by means of a specially coded numeric keyboard and thermal printer (and CRT on ASTRA 8). The operator can program a sequence of up to 80 samples with specimen identification number and appropriate tests for each. Routine runs can be interrupted to perform STAT analyses. The sampling rate varies from 70 to 85 samples per hour with a turnaround time of less than 2 min. The user can define the expected or reference ranges for each test. Ranges of acceptable results for three quality control materials can be defined by the user for on-line quality control monitoring. Results for patient or quality control samples that are above or below the defined limits are flagged as "LO" or "HI."

The analytical components of the instrument include a 40-cup sample tray, dual sample probe with transport, a sampling and wash module, and the analytical modules. A conductivity measurement is made between the two probes while sampling to check for adequate volume in the sample cup. The probes are then rinsed off and sequentially inserted into the measuring cuvets (two cuvets at a time) until sample aliquots for all tests requested have been delivered. Both the sampling and reagent dispensing units use peristaltic pumps. Preset, non-changeable sample volumes ranging from 8 to 50 μliter are used for the various tests. Reagent is delivered in excess into the measuring cell, and the excess is removed through a port located on the side of the cell.

A two-point calibration is automatically performed at a fixed 50-min time interval. The operator may request more frequent calibrations. The drift from the previous calibration is printed out. All channels, or just those specifically requested, can be calibrated at one time. Since the user cannot redefine calibration set points, the user is locked into using the particular calibrating solutions made by Beckman or some facsimile thereof. If an in-house preparation or other commercial source is used, it must contain all analytes being measured on the ASTRA at the specified concentrations.

An important feature of the ASTRA is its self-diagnostic capability. Error messages are automatically displayed or printed, and results flagged when errors are detected in the measuring process. To assist in identifying and troubleshooting errors, the operator can call up diagnostic routines for each module. These routines step through the entire measurement process from sampling to detection in a stepwise fashion under test-simulated conditions. The operator can observe proper function in addition to the instrument sensors. Raw signal levels from the detector provide important diagnostic information.

The analytical performance of the ASTRA has been reported by Finley *et al.* (1978), Truchaud *et al.* (1980), and Hartmann and Fillbach (1980). Fievet *et al.* (1980) studied the response of the sodium (glass) and the potassium (valinomycin) electrodes. Over the expected ranges for urine specimens, 0–200 mmol/liter, sodium response was described by a third order polynomial and potassium by a fourth order polynomial. This is in contrast to the typical logarithmic Nernstian behavior.

B. Blood Gas in Situ Analyzers

Blood gas analyzers typically perform two, three, or four different measurements on a single blood specimen by means of electrodes and cuvet(s) mounted in series in a sample line. Previously, the analytical performance of blood gas analyzers had been extremely dependent on operator skill. Microprocessor control of the instrument in all aspects should reduce this dependence and improve operator-to-operator consistency. The IL System 1303 (Instrumentation Laboratory, Lexington, Massachusetts 02173) is an example of a microprocessor-controlled blood gas analyzer.

The microprocessor's control of the instrument functions is supported by prompting messages and diagnostic displays on the CRT and by operator instructions transmitted through the coded numeric keyboard. A tape printer is used to provide hard copy of all test results and calculations. The flexibility in operation due to the microprocessor is further enhanced by the variety of sampling techniques. Aspiration or injection from a syringe, aspiration from capillary tubes, and aspiration from quality control vials are all accomplished through the same sampling probe for sample volumes of either 120 or 65 μliter. The reduction in sample volume and flexibility in sampling techniques are major improvements over earlier models in the series of IL blood gas analyzers.

The sample line includes a preheater and the three electrodes mounted in series in a solid heating block maintained at 37°C. The electrode measuring chambers are visible to the operator (traditional in all IL analyzers) for detection of clots and bubbles. An in-line valve is located between the pH electrode and the gas electrodes to facilitate simultaneous calibration with gases for the Pco_2 and Po_2 electrodes and buffers for the pH electrode. Continuous one-point calibrations of the three electrodes and intermittent two-point calibrations (at user definable intervals) are performed automatically. During the measuring process, the microprocessor monitors (and continuously displays on the CRT) the electrode signals for endpoint or steady state values for up to 2 min. Unstable signals at 2 min are flagged and suggest electrode cleaning or replacement, or an unstable sample due to the presence of air bubbles.

Data such as the specimen identification number, the hemoglobin value, the patient's temperature and FIO2 (fraction of inspired oxygen) can be entered into the keyboard. The hemoglobin value enables the calculation of acid–base parameters such as base excess, bicarbonate and total CO_2 from the measured pH, Pco_2 and Po_2. In vitro temperature corrections can be determined using algorithms stored in memory.

The microprocessor can also be connected to the IL 282 Co-oximeter™ for automatic transmission of the hemoglobin parameters measured by the co-oximeter. In this case, the measured hemoglobin is used automatically to calculate the acid–base parameters. The combined data can be printed out on one report or transmitted together to a central laboratory computer. Jones (1980) has provided preliminary data on the analytical performance of the IL 1303.

As has been noted, the microprocessors in blood gas analyzers have the capability of calculating blood gas parameters for in vitro changes in temperature. The validity of performing such adjustments is open to question (Blume, 1978; Hansen and Sue, 1980). At present there are few data available regarding the reference intervals or normal values of blood gas parameters at temperatures other than 37°C. This prevents proper interpretation of results that are reported for some other temperature. Although the laboratory and the clinical instrumentation is sophisticated enough to perform these in vitro adjustments, care must be taken to provide results that are meaningful to the physician.

Although the microprocessor has improved laboratory operations, made difficult testing simple, shortened analysis time, and improved reliability, the key driving force for the development of new methodologies and innovations must be medical need (Astrup, 1979).

References

Abbott Laboratories (1980). "Abbott VP Bichromatic Analyzer Operations Manual." Abbott Laboratories, Diagnostics Division, Dallas, Texas.

Anderson, N. G. (1969). *Anal. Biochem.* **28**, 545–562.

Astrup, P. (1979). *Ann. Clin. Biochem.* **16**, 338–342.

Baadenhuijsen, H., and Seuren-Jacobs, H. E. H. (1979). *Clin. Chem.* **25**, 443–445.

Bagshawe, K. D. (1978). *In* "Automated Immunoanalysis by Radioimmunoassay" (R. F. Ritchie, ed.), pp. 541–548. Dekker, New York.

Bandi, Z. L., Fuller, J. B., Bee, D. E., and James, G. P. (1981a). *Clin. Chem.* **27,** 27–34.

Bandi, Z. L., Fuller, J. B., Bee, D. E., and James, G. P. (1981b). *Clin. Chem.* **27,** 480–485.

Betteridge, D. (1978). *Anal. Chem.* **50,** 832A–846A.

Blake, B. F., Chapman, W. A., Lloyd, G. A., and Teisteeg, G. E. (1980). *Clin. Chem.* **26,** 974.

Blume, P. (1978). *Am. J. Clin. Pathol.* **70,** 440–441.

Bowie, L. J. (1980). "Automated Instrumentation for Radioimmunoassay." CRC Press, Boca Raton, Florida.

Brooker, G., and Murad, F. (1980). *Clin. Chem.* **26,** 1738–1740.

Brooker, G., Terasaki, W. L., and Price, M. G. (1976). *Science* **194,** 270–276.

Button, K., and Gambino, S. Z. (1977). *Clin. Chem.* **23,** 1151.

Carey, R. N., Eggert, A. A., Quam, E. F., Feldbruegge, D. H., and Westgard, J. O. (1977). *Adv. Autom. Anal., Technicon Int. Congr., 7th, 1976* pp. 223–226.

Carlyle, J. E., McLelland, A. S., and Fleck, A. (1973). *Clin. Chim. Acta* **46,** 235–241.

Carter, O. L., and Grimmet, M. G. (1976). *Clin. Chem.* **22,** 1165.

Carter, O. L., and Grimmett, M. G. (1977). *Clin. Chem.* **23,** 1152.

Cate, J. C., IV, Hendrick, R., Taylor, M., and McGlothlin, C. D. (1980). *Clin. Chem.* **26,** 266–270.

Chen, I. W. (1980a). *Ligand Rev.* **2**(2), 46–50.

Chen, I. W. (1980b). *Ligand Rev.* **2**(3), 46–48.

Chen, I. W., Maxon, H. R., Heminger, L. A., Ellis, K. S., and Volle, C. P. (1980c). *J. Nucl. Med.* **21,** 1162–1168.

Cohen, E., and Stern, M. (1977). *Adv. Autom. Anal., Technicon Int. Congr., 7th, 1976* pp. 232–234.

Curme, H. G., Columbus, R. L., Dapper, G. M., Eder, T. W., Fellows, W. D., Figueras, J., Glover, C. P., Goffe, C. A., Hill, D. E., Lawton, W. H., Muka, E. J., Pinney, J. E., Rand, R. N., Sanford, K. J., and Wu, T. W. (1978). *Clin. Chem.* **24,** 1335–1342.

Curme, H. G., Babaoglu, K., Babb, B. E., Battaglia, C. J., Beavers, D. J., Bogdanowicz, M. J., Chang, J. C., Daniel, D. S., Kim, S. H., Kissel, T. R., Sandifer, J. R., Schnipelsky, P. N., Searle, R., Secord, D. S., and Spayd, R. W. (1979). *Clin. Chem.* **25,** 1115.

Daniel, D. S., Babb, B. E., Battaglia, C. J., Bogdanowicz, M. J., Chang, J. C., Kim, S. H., Kissel, T. R., Sandifer, J. R., Schnipelsky, P. N., Searle, R., and Secord, D. S. (1980). *Clin. Chem.* **26,** 990.

Dass, S., Bowen, N. L., and Bagshawe, K. D. (1980). *Clin. Chem.* **26,** 1583–1587.

de Graeve, J. S., Andrieu, N., Valdiquie, P., and Fichant, G. (1981). *Clin. Chem.* **27,** 337–338.

de Haan, J. B. (1979). *In* "Topics in Automatic Chemical Analysis" (J. K. Foreman and P. B. Stockwell, eds.), Vol. I, pp. 208–236. Wiley, New York.

Dolan, S. J., Van der Wal, Sj., Bannister, S. J., and Snyder, L. R., (1980). *Clin. Chem.* **26,** 871–880.

Eggert, A. A., and Westgard, J. O. (1975). *Clin. Chem.* **21,** 1005.

Eisenweiner, H. G., and Keller, M. (1979). *Clin. Chem.* **25,** 117–121.

Elser, R. C. and Garver, C. G. (1981). *Clin. Chem.* **27,** 325–327.

Evenson, M. A., Hicks, G. P., and Thiers, R. E. (1970). *Clin. Chem.* **16,** 606–611.

Fievet, P., Truchaud, A., Hersaut, J., and Glikmanas, G. (1980). *Clin. Chem.* **26,** 138–139.

Finley, P. R., Williams, R. J., Lichti, D. A., and Thies, A. C. (1978). *Clin. Chem.* **24,** 2125–2131.

Forrest, G. (1977). *Adv. Autom. Anal., Technicon Int. Congr., 7th, 1976* pp. 241–246.

Fu, P., Witte, D., Brown, L. F., Neu, B., and Lubran, M. M. (1979). *Clin. Chem.* **25,** 1116.

Furman, W. B. (1976). "Continuous Flow Analysis: Theory and Practise." Dekker, New York.

Garber, C. C., and Miller, R. C. (1980). *Clin. Chem.* **26,** 989.

Garber, C. C., Feldbruegge, D., Miller, R. C., and Carey, R. N. (1978). *Clin. Chem.* **24,** 1186–1190.

Garber, C. C., Westgard, J. O., Milz, L., and Larson, F. C. (1979). *Clin. Chem.* **25,** 1730–1738.

Garber, C. C., Feldbruegge, D. H., and Hoessel, M. (1981). *Clin. Chem.* **27,** 614–619.

Habig, R. L., Schein, B. W., Walters, L., and Thiers, R. E. (1969). *Clin. Chem.* **15,** 1045–1055.

Hansen, J. E., and Sue, D. Y. (1980). *N. Engl. J. Med.* **303,** 341.

Hartmann, A. E., and Fillbach, J. R. (1980). *Am. J. Clin. Pathol.* **74,** 275–281.

Hills, L. P., and Tiffany, T. O. (1980). *Clin. Chem.* **26,** 1459–1466.

Horvath, C., and Pedersen, H. (1977). *Adv. Autom. Anal., Technicon Int. Congr., 7th, 1976* pp. 86–95.

Johnson, E. G., Sturgis, B. E., and Stonecypher, T. E. (1976). *Clin. Chem.* **22,** 1164.

Jones R. J. (1980). *Clin. Chem.* **26,** 1032.

Karcher, R. E., and Foreback, C. C. (1977). *Adv. Autom. Anal., Technicon Int. Congr., 7th, 1976* pp. 191–196.

Karmel, R., Landon, J., and Forrest, G. (1980). *Clin. Chem.* **26,** 97–100.

Kim, S. H., Babb, B. E., Bogdanowicz, M. J., Chang, J. C., Daniel, D. S., Kissel, T. R., Pipel, M. W., Sandifer, J. R., Schnipelsky, P. N., Searle, R., Spayd, R. W., and Steele, T. J. (1980). *Clin. Chem.* **26,** 990.

Lawn, W. G., Grimmett, M. G., and Carter, O. L. (1977). *Clin. Chem.* **23,** 1151.

Leon, L. P., Sonsur, M., Snyder, L. R., and Horvath, C. (1977). *Clin. Chem.* **23,** 1556–1562.

Margoshes, M. (1977). *Anal. Chem.* **49,** 17–19.

Miller, D. T., Martin, S. H., Lehane, D. P., and Rossi, R. J. (1980). *Clin. Chem. (Winston-Salem, N.C.)* **26,** 1073.

Mitchell, F. L. (1980). *In* "Centrifugal Analysis in Clinical Chemistry" (C. P. Price and K. Spencer, eds.), pp. 311–323. Praeger, New York.

National Committee for Clinical Laboratory Standards (NCCLS) (1976). "Protocol for Establishing the Precision and Accuracy of Automated Analytic Systems," PSEP-1. NCCLS, Villanova, Pennsylvania.

National Committee for Clinical Laboratory Standards (NCCLS) (1979). "Protocol for Establishing Performance Claims for Clinical Chemical Methods: Introduction and Performance Check Experiment, Replication Experiment, and Comparison of Methods Experiment," EP2-P, EP3-P, and EP4-P. NCCLS, Villanova, Pennsylvania.

Neeley, W. E., Wardlow, S., and Swinnen, M. E. T. (1974). *Clin. Chem.* **20,** 78–80.

O'Leary, N., and Duggan, P. F. (1980). *Clin. Chem.* **26,** 793.

Painter, K., and Hasler, M. J. (1976). *Clin. Chem.* **22,** 1164.

Parker, N. C., and Cross, R. E. (1980). *Clin. Chem.* **26,** 1074.

Pearson, K. W., Smith, R. E., Mitchell, A. R., and Biasel, E. R. (1981). *Clin. Chem.* **27,** 256–262.

Percy-Robb, I. W., Simpson, D., Taylor, R. H., and Whitby, L. G. (1978). *Clin. Chem.* **24,** 146–148.

Piknosh, W., Goldman, S. C., and Wheaton, B. A. (1979). *Clin. Chem.* **25,** 1107.

Powers, D. M., Rand, R. N., and Brody, B. B. (1980). *Clin. Chem.* **26,** 991.

Quick, R. F., Thew, C. A., and Thiers, R. E. (1980). *Clin. Chem.* **26,** 1014.

Rand, R. N., Kussee, D. G., and O'Brien, P. A. (1980). *Clin. Chem.* **26,** 974.

Ranger, C. B. (1981). *Anal. Chem.* **53,** 20A–32A.

Reese, M. G., and Johnson, L. R. (1978). *Clin. Chem.* **24,** 342–344.

Renoe, B. W., Stewart, K. K., Beecher, G. R., Wills, M. R., and Savory, J. (1980). *Clin. Chem.* **26,** 331–334.

Robertson, E. A., and Young, D. S. (1977). *Adv. Autom. Anal., Technicon Int. Congr., 7th, 1976* pp. 204–210.

Robertson, E. A., Van Steirteghem, A. C., and Young, D. S. (1979). *Clin. Chem.* **25,** 1121.

Ross, J. W., Martin, D. G., and Moore, T. D. (1980). *Am. J. Clin. Pathol.* **74**, 521–530.

Rush, R. L. and Nabb, D. P. (1977). *Adv. Autom. Anal., Technicon Int. Congr., 7th, 1976* pp. 197–203.

Savory, J., and Cross, R. E. (1978). "Methods for the Centrifugal Analyzer." Am. Assoc. Clin. Chem., Winston-Salem, North Carolina.

Schlabach, T. D., Fulton, J. A., Mockridge, P. B., and Toren, E. C. (1979). *Clin. Chem.* **25**, 1600–1607.

Schnipelsky, P. N., Glover, C. P., Akubowicz, R. F., and Larson, D. E. (1979). *Clin. Chem.* **25**, 1115.

Schwartz, M. K. (1978). *In* "Recent Advances in Clinical Biochemistry" (Alberti, K. G. M. M., ed.), pp. 229–253. Churchill-Livingstone, New York.

Schwartz, M. K., Bethune, V. G., and Fleischer, M. (1974). *Clin. Chem.* **20**, 1062–1070.

Simkowski, K. W., King, R., and Foreback, C. C. (1980). *Clin. Chem.* **26**, 1074.

Skeggs, L. T. (1957). *Am. J. Clin. Pathol.* **28**, 311–322.

Snook, M., Renshaw, A., Ridcout, J. M., Wright, D. J., Baker, J., and Dickins, J. (1979). *J. Autom. Chem.* **1**, 72–77.

Snyder, L. R. (1976). *J. Chromatogr.* **125**, 287–306.

Snyder, L. R. (1977). *Adv. Autom. Anal., Technicon Int. Congr., 7th, 1976* pp. 76–81.

Snyder, L. R., and Alder, H. J. (1976a). *Anal. Chem.* **48**, 1017–1022.

Snyder, L. R., and Alder, H. J. (1976b). *Anal. Chem.* **48**, 1022–1027.

Spayd, R. W., Bruschi, B., Burdick, B. A., Dappen, G. M., Eikenberry, J. N., Esders, T. W., Figueras, J., Goddhue, C. T., LaRossa, D. D., Nelson, R. W., Rand, R. N., and Wu, T. W. (1978). *Clin. Chem.* **24**, 1343–1350.

Spencer, K. (1976). *Ann. Clin. Biochem.* **13**, 438–448.

Squibb, E. R. & Sons, Inc. (1980). "Operations Manual of the Gammaflo R Automated Assay System." Princeton, New Jersey.

Stamper, R., and Robertshaw, D. M. (1980). *Clin. Chem.* **26**, 778–780.

Technicon Instruments Corp. (1980). "Product Labeling, Technicon STAR System," Tech. Publ. No. UA 80-430-00. Tarrytown, New York.

Thiers, R. E., Cole, R. R., and Kirsch, W. J. (1967). *Clin. Chem.* **13**, 451–467.

Thiers, R. E., Meyn, J., and Wilderman, R. F. (1970). *Clin. Chem.* **16**, 832–839.

Thiers, R. E., Reed, A. H., and Delander, K. (1971). *Clin. Chem.* **17**, 42–48.

Truchaud, A., Hersant, J., Glikmanas, G., Fievet, P., and Dubois, O. (1980). *Clin. Chem.* **26**, 139–141.

Valdes, R., Jr., Savory, G., Bruns, D., Renoe, R., Savory, J., and Wills, M. R. (1979). *Clin. Chem.* **25**, 1254–1258.

Walker, W. H. C. (1976). *In* "Continuous Flow Analysis" (W. B. Furman, ed.), pp. 207–225. Dekker, New York.

Walker, W. H. C. (1977). *Adv. Autom. Anal., Technicon Int. Congr., 7th, 1976* pp. 82–85.

Walker, W. H. C., and Andrew, K. (1974). *Clin. Chim. Acta* **57**, 181–185.

Walker, W. H. C., Sheperdson, J. C., and McGowan, G. K. (1971). *Clin. Chim. Acta* **35**, 455–460.

Walmsley, T. A., Abernathy, M. H., and Fowler, R. T. (1980). *Clin. Chem.* **26**, 530–531.

Warren, K., Kubasik, N. P., Brody, B. B., Sine, H. E., and D'Souza, J. P. (1980). *Clin. Chem.* **26**, 133–137.

Westgard, J. O., and Groth, T. (1979). *Clin. Chem.* **25**, 863–869.

Westgard, J. O., and Lahmeyer, B. L. (1972). *Clin. Chem.* **18**, 340–348.

Westgard, J. O., Carey, R. N., Feldbruegge, D. H., and Jenkins, L. M. (1976). *Clin. Chem.* **22**, 489–496.

Westgard, J. O., Groth, T., Aronsson, T., Falk, H., and de Verdier, C. (1977a). *Clin. Chem.* **23**, 1857–1867.

Westgard, J. O., Groth, T., Aronsson, T., and de Verdier, C. (1977b). *Clin. Chem.* **23,** 1881–1887.

Westgard, J. O., Falk, H., and Groth, T. (1979). *Clin. Chem.* **25,** 394–400.

Williams, F. C., and Clapper, F. R. (1953). *J. Opt. Soc. Am.* **43,** 595–599.

Wu, T. W., Lo, D. H., Dappen, G. M., and Spayd, R. W. (1980). *Clin. Chem.* **26,** 990.

Yalow, R. S., and Berson, S. A. (1959). *Nature (London)* **184,** 1648–1699.

Zborowski, G., and Woo, C. (1977). *Adv. Autom. Anal., Technicon Int. Congr., 7th, 1976* pp. 235–240.

8

Continuous Automated Analysis of Gases and Particulates in the Pulp and Paper Industry

T. L. C. DE SOUZA

Pulp and Paper Research Institute of Canada
Pointe Claire, Quebec
Canada

I. Introduction

The pulp and paper industry inadvertently generates pollution in four basic forms, namely, emission of malodorous gases, particulates, pollution of large volumes of water used in the various processes, and noise in the work place. This chapter deals principally with automation or more accurately, the use of continuous analyzers in the industry to monitor emissions of gases and particulates.

Testing of malodorous gases and particulate matter emitted from pulp and paper mills can be performed by both batch and continuous sampling methods. Batch testing provides an average concentration value for a given time period. Continuous monitoring provides a record of instantaneous concentration values over a prolonged time interval to determine whether there is compliance with air pollution regulations and helps to monitor the operations of process equipment. Continuous monitoring instruments normally require a higher capital cost than batch sampling equipment, but once these systems are placed in operation and maintained by competent and trained personnel, the manpower requirements are lower.

Identification and quantitation of gaseous and particulate materials in flue gases must be performed in order to be able to stay in compliance with existing and proposed pollutant emission standards and also to be able to make an inventory of material losses. Developing methods for direct and accurate measurement of air pollutants such as total reduced sulfur (TRS) compounds, oxides of sulfur, oxides of nitrogen, and particulate matter is important because of increasingly stringent emission standards at all levels of government.

Successful operation of continuous or automatic monitoring systems for measuring emissions of gases and particulates requires instrumentation that is accurate, reliable, stable, simple to operate, low in maintenance requirements, and subject to minimal interferences. In designing and operating an automatic monitor, sample preconditioning, sample handling, type of detector, and data reducing and reporting systems should be carefully considered.

II. Monitoring of Gases

The principal types of gaseous pollutants emitted from pulp and paper industry sources that may require continuous monitoring are malodorous reduced sulfur compounds, such as hydrogen sulfide (H_2S), methyl mercaptan (CH_3SH), dimethyl sulfide [$(CH_3)_2S$], and dimethyl disulfide [$(CH_3)_2S_2$], commonly classified together as TRS compounds. Oxides of sulfur such as sulfur dioxide (SO_2) and sulfur trioxide (SO_3), commonly referred to as (SO_x), oxides of nitrogen such as nitric oxide (NO) and nitrogen dioxide (NO_2), commonly referred to as (NO_x), oxygen (O_2) and other sulfur-free compounds are also monitored.

Both H_2S and organic mercaptans and sulfides are extremely odorous (Wilby, 1969), detectable at concentrations of only a few parts per billion. Thus odor control is one of the principal air pollution problems in a kraft pulp mill. Potential points of emissions of reduced sulfur gas to the atmosphere (Fig. 1) include the digester, washer hood, multiple-effect evaporator, direct contact evaporator, recovery furnace, smelt dissolving tank, slaker vent, black liquor oxidation tank, wastewater treatment operations, and lime kiln.

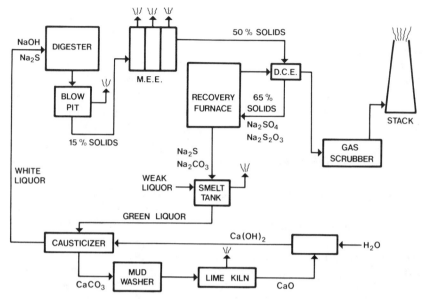

Fig. 1. Schematic diagram of a kraft chemical recovery system.

Oxides of sulfur can be emitted in varying quantities in kraft and sulfite chemical recovery systems. The major source of sulfur dioxide emissions is from the chemical recovery furnace, because of the combustion of sulfur-containing spent liquor. Under certain conditions, somewhat similar quantities of sulfur trioxide can also be released to the atmosphere (Maksimov *et al.,* 1965), particularly when heavy fuel oil is used as an auxiliary fuel. Lesser quantities of SO_2 can also be released from the lime kiln, smelt dissolving tank, and other kraft mill sources.

Oxides of nitrogen can be formed in any fuel combustion process by the reaction between oxygen and nitrogen at high temperatures. The major constituent formed is nitric oxide, a small portion of which can be oxidized to form nitrogen dioxide. Nitrogen oxide emissions from kraft pulp mill process sources, such as the recovery furnace and lime kiln, are normally lower than those for most other fuel combustion processes. This is primarily because black liquor and lime mud when burnt give off large quantities of steam, which acts as a heat sink to suppress the flame temperature. Larger quantities of oxides of nitrogen can be formed, however, when auxiliary fuels such as natural gas or fuel oil are burnt.

Non-sulfur-containing organic compounds can also be emitted in varying quantities from several different kraft and sulfite pulp mill process sources. The major types of materials that can be released to the atmosphere include terpenes, hydrocarbons, alcohols, phenols, and other organic compounds liberated from

wood pulp. The primary significance of these compounds, particularly the terpenes, is that they may act either directly as odorant gases or as liquid particulate
carriers for odorous sulfur gas molecules. The olefinic hydrocarbons or terpenes
could undergo photochemical reactions in polluted atmospheres. Major process
variables that affect emissions of these compounds to the atmosphere include the
wood species being pulped, temperature used in the process, the volatility of the
organic compounds released, the type of paper coating material used, and the
type and effectiveness of any air pollution control device employed.

Emissions from the sulfite pulping process are principally SO_2 and particulate
matter. In special cases of burning alkaline sulfite liquor in recovery furnaces
under reducing conditions, H_2S emissions may also occur. Otherwise, there are
practically no organic reduced sulfur compounds produced in the sulfite process.
The main sources of SO_2 emissions are the digester blow pits, multiple-effect
evaporators, and liquor burning or chemical recovery systems. Minor process
sources include pulp washers and acid preparation plants. Nitrogen oxides are
emitted from various combustion sources, particularly from the recovery furnaces of ammonium-based mills.

A. General Sampling Systems

The purpose of a sample handling and conditioning system is to remove the
sample from the flue gas duct and transfer it to the analytical instrument for
subsequent analysis without changing the concentration or character of the constituents to be measured. The major elements of sample conditioning system
(Cooper and Rossano, 1971) are the sample probe, particulate and moisture
removal devices, transfer tubing, and a prime mover. Generally, the construction
materials and the sample conditioning system used for TRS compounds will be
satisfactory for compounds like SO_2, NO_x, O_2, CO_2, etc., as well. Hence, these
topics will not be discussed again when the above-mentioned gaseous compounds are dealt with.

B. Monitoring of TRS—Materials of Construction

Any material that can lead to adsorption or chemical reaction with sulfur
compounds should be avoided. Teflon, glass, and Carpenter stainless steel are
found by a number of researchers to be suitable materials for sample lines and
interfaces. Some papers and publications (e.g., Nader, 1973; de Souza et al.,
1975a; National Information Service, 1976) give useful information and advice
on maintaining sample integrity.

C. Sample Probe

The probe is usually made of stainless steel or ceramic and is located within the stack to facilitate removal of a portion of the moving gas stream for channeling into the sample conditioning system. Larger particles are held back in the duct by a plug of glass wool (Blosser *et al.*, 1968) inserted into the open end of the probe for manual operation. For automatic sampling, the probe is equipped with a sintered head of stainless steel or ceramic, and a compressed air blowback feature (Thoen *et al.*, 1969a) is actuated with a solenoid valve and a timing device.

D. Sample Conditioning

The function of conditioning devices is usually two-fold, viz., to prevent condensation of water vapor in the sample stream and to remove fine particulate matter that can, with time, coat the walls of the transport system and foul the insides of the analytical instrument.

Transferring sulfur compounds from the flue gas duct to the detector without changing the concentration or character of the sample gas is complicated by the presence of large quantities of water vapor and particulate matter in the flue gas at high temperatures. In measuring TRS, it is sometimes necessary to first remove sulfur dioxide from combustion source gases by the use of a liquid scrubber containing a preconditioned solution of potassium acid phthalate or a mixture of citric acid and potassium citrate, located immediately downstream of the probe. The scrubber, which can be either continuous flow or batch type, selectively removes sulfur dioxide from the flue gas, condenses excess water vapor, removes additional particulate matter, and cools the gas stream from stack to ambient temperature conditions.

In certain gas conditioning systems (Fig. 2), after the removal of particulates and sulfur dioxide, the sample gas is passed through a combustion furnace placed upstream of the detector (Canfield, 1971) and at a temperature of approximately $825 \pm 25°C$ in order to convert reduced sulfur compounds to SO_2 in the presence of O_2, usually found in flue gases. This conversion or oxidation makes it feasible to use SO_2-sensitive detection methods, such as flame photometry, ultraviolet spectrophotometry, and electrochemical analysis. An additional advantage of precombustion is that it converts potentially interfering organic compounds, such as terpenes and olefinic and aromatic hydrocarbons to nonreactive carbon dioxide and water. The major drawback of the conversion step is that nonodoriferous compounds of sulfur, e.g., carbonyl sulfide (COS) often emitted under overloaded furnace operating conditions (Thoen and Nicholson, 1970; Bhatia *et al.*, 1975), will also be converted to SO_2 and erroneously reported as TRS.

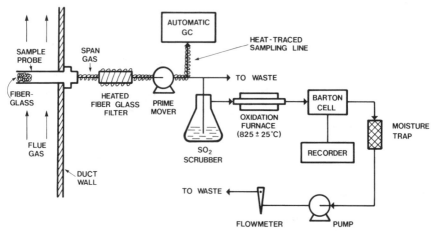

Fig. 2.   Sample gas conditioning train.

1.   PRIME MOVER

The prime mover is a device to transport the sample gas from the stack duct to the analytical instrument in the least possible time. It can be located either upstream or downstream of the detector depending upon the system used. A positive displacement pump, upstream of the detection unit, is particularly desirable if a considerable volume of stack gas is removed for quick transfer to the detector or if the latter must be operated under positive pressure. Potential problems with pumps are particulate plugging, moisture condensation, and corrosion. Air-, water-, or steam-operated aspirators can also be used in place of the pump.

2.   SAMPLING LINES

Transport tubing should be made of an inert material to avoid possible sample losses by physical adsorption or chemical reaction with the constituents of the sample gas. Electrically heated Teflon tubing is found to be the best for this purpose and can be purchased commercially. Materials of Tygon, steel, copper, rubber, etc., are unfit for the transporation of reactive sulfur compounds. The diameter of the tubing should be such as to offer minimum back pressure but at the same time minimum holdup in the tubing as well. This can usually be attained in tubing of 0.64–1.27 cm internal diameter.

3.   PARTICULATE REMOVAL SYSTEMS

Coarse particulates can be removed in situ in the flue gas duct by one of the following ways:

(i)   by the use of a plug of fiber glass at one end of a probe made of stainless steel tube;

(ii)   by the use of an air blowback system employing a sintered stainless steel or ceramic head placed in a stainless steel tube;

(iii)   by the use of an "inertial" filter (Metallurgical Corporation, 1975) that utilises the principle of particulates segregating to the middle of a fast moving gas stream; and

(iv)   a truly continuous sample probe with a blowback made of stainless steel developed by de Souza (1982).

Fine particulates can be filtered off by the use of small-diameter pore-size fiberglass filters placed downstream of the coarse filter.

4.   DEALING WITH WATER VAPOR

Condensation of water vapor present in the sample gas can be prevented in any one of the following ways: (1) heating the transport tubing above the dew point temperature of the sample gas, (2) diluting the sample gas stream with N_2 or air (when oxidation is unimportant), and (3) using a semipermeable membrane to specifically separate water vapor from other gas constituents. Each one of these methods is well documented in the literature. Careful consideration must be given to their advantages and disadvantages before any one of these techniques is adopted. The alternative is to allow condensation to occur in a catchpot containing some acid, such as H_3PO_4, which helps to release any sulfur gas, such as, H_2S that might be absorbed by the condensate back into the gaseous phase.

E.   Gas Detection Systems

Continuous or automated monitoring of reduced sulfur compounds in kraft pulp mills is usually achieved with flame photometry, ultraviolet spectroscopy, coulometric titration, or electrochemical sensing.

1.   ANALYZERS WITH FLAME PHOTOMETRIC DETECTORS

The principle for the detection of sulfur and phosphorous compounds, discovered and revealed by Dragerwerk and Drager (1962), was first applied to produce a flame photometric detector (FPD) for gas chromatography in 1966 by Brody and Chaney. The detector is based upon the burning, in a hydrogen-rich flame, of compounds of phosphorus and sulfur which emit characteristic green and blue flames, respectively. Narrow bandpass interference filters are utilized for spectral isolation of phosphorous emission at 526 nm and sulfur emission at

394 nm. Interferences produced by other organic compounds and carbon dioxide in the flame are eliminated by an optical arrangement whereby only the interference-free tip of a normal flame envelope is viewed by the photomultiplier tube. This is the single-flame detector. A second detector, which is an improved version of this detector, utilises two flames (Gangwal and Wagoner, 1979), one on the top of the other. The lower flame is used to decompose and eliminate interference due to organic compounds and to oxidize all the sulfur compounds to SO_2, which is then the sole compound to be excited and monitored in the second flame. The advantage of the single-flame detector is its higher sensitivity to sulfur compounds, whereas the dual-flame detector is excellent for those samples that have backgrounds of high organic contents (Ferguson and Luke, 1979) and is reported to offer broader linear concentration ranges with uniform response to all sulfur species.

The flame photometric detector responds to most sulfur gases and on a log–log scale gives linear relationships in the near-zero to 1 ppm concentration range. Two types of instruments with the FPD detector are available, viz. the nonseparating and the separating type. In the former, separation of the individual gas constituents from a mixture is not attempted, whereas in the latter, it is. Instruments of the nonseparating type are limited to measuring directly and continuously the total sulfur or "TRS" (including other nonodoriferous sulfur compounds),* which, in any case, would require a SO_2 scrubber ahead of the detector. An example of the separating type is the gas chromatograph.

2. GAS CHROMATOGRAPHIC ANALYZERS

The separation of a sample gas mixture into individual components is most useful in distinguishing between odoriferous and nonodoriferous sulfur compounds under analysis. Also, knowing the type of sulfur compounds emitted, e.g., organic or inorganic, helps the operator of an odor abatement device, such as a wet scrubber, to use the best available technology—for example, the correct reagent solution in the wet scrubbing of the gases encountered. This separation of the sample gas mixture is achieved with gas chromatographic instruments equipped with packed columns with a high surface-to-volume ratio. A fine solid packing material or a liquid dispersed on an inert solid offers the high surface area needed for the separation. Long capillary tubes, without any packing (Blomberg, 1976), can also be used. The separation of the gas mixture is achieved by exploiting the differences in relative affinities of these constituents for the packing material or liquid phase used in the separating column. These differences cause the various constituents of the mixture to pass through the column at different rates. An inert carrier gas moves the sample through the column. Each

*Meloy Laboratories Inc., Springfield, Virginia 22151.

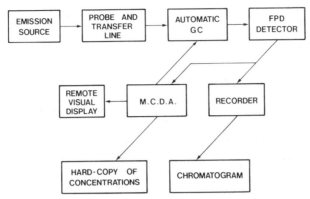

Fig. 3. Design of the PPRIC automatic GC monitor for the measurement of TRS compounds.

component of the sample gas mixture requires a different amount of time to pass through the column, the weakly adsorbed components being the first to emerge. A suitable sensor can then measure the concentration of each compound as it leaves the column. The separation process is highly temperature-sensitive and thus requires that most components of the gas chromatograph (GC) be housed in a temperature-controlled oven.

Gas chromatographic instruments, until now, were used by the pulp and paper industry principally as batch-type laboratory analyzers or as ambient monitors. However, automatic GC monitors are now available. The one developed at the Pulp and Paper Research Institute of Canada (PPRIC) by de Souza *et al.* (1978) separates H_2S and COS, analyzes all six sulfur components usually found in emissions of kraft recovery furnaces from a single sample injection without dilution, and has a cycle time of only 10 min. In this instrument, the sampling, analytical procedure, data collection, data processing, and printing and/or display of the results directly in parts-per-million concentrations are fully automated with all functions of the system being controlled by a microcomputer. Figure 3 illustrates the overall conceptual design of the GC monitor. The sample gas is withdrawn from the source duct and transported to the GC through heat-traced Teflon tubing after the separation of particulate matter. Two valves of Carpenter stainless steel (10-port and 4-port) are used. Sampling and injecting the sample gas into the gas chromatographic column are performed as shown in Fig. 4. The individually separated gas constituents (Fig. 5) eluting from three specially treated Porapak QS separating columns maintained at isothermal temperatures (de Souza *et al.*, 1975b) are burnt in the FPD burner, where the characteristic blue light of the active sulfur compounds is converted into electrical signals that are monitored on a strip-chart recorder. A microcomputer data acquisition (MCDA) system converts the detector signals to concentration units (ppm) by relating the signals to those obtained with standard gas mixtures, which information is stored

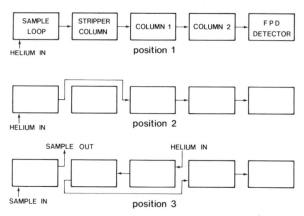

Fig. 4. Diagram of the separating procedure.

in memory. The computer is built with a central processor that has an 8-bit word length with a 64,000 maximum word addressability and 78 basic instructions. The memory section is capable of storing 8000 words, whereas the I/O Logic and Interface is a program-dependent circuit.

The GC signal obtained is made available to the microcomputer in both the

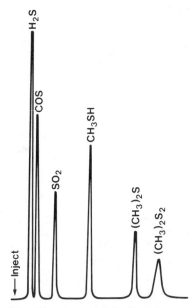

Fig. 5. Chromatogram of six sulfur compounds in kraft pulping operations. H_2S = 40 ppm; COS = 30 ppm; SO_2 = 25 ppm; CH_3SH = 25 ppm; $(CH_3)_2S$ = 18 ppm; $(CH_3)_2S_2$ = 23 ppm.

TABLE I

Typical Printout after Analysis

GAS	START[a]	MAX[a]	END[a]	AREA	PPM
H2S	48	55	66	15414	9.1
COS	66	74	115	09299	1.2
SO2	115	133	230	22549	16.8
RSH[b]	230	270	397	42926	10.6
R2S	397	428	481	11053	2.1
R2S2	481	536	590	06423	0.7
TRS					23.2

[a] Entries are times, in seconds, from time of injection.
[b] $R \equiv CH_3$.

direct and linearized forms. The results obtained are printed out as individual sulfur compounds with their respective concentrations followed by the composite TRS number (Table I). The instrument has been modified to use a single temperature programmed, specially treated Porapak QS separating column (de Souza *et al.*, 1975b) and is now licensed to be commercially produced by Western Research and Development, Calgary, Alberta, Canada. To date, it is one of only three fully automatic GC instruments built and dedicated specifically for quantitative measurements of individual as well as total reduced sulfur compounds from kraft mill emissions. The other two are by Tracor Analytical Instruments and Bendix Environmental and Process Instrumentation Division.

A gas chromatograph can be said to be made up of a number of small sections, such as follows.

a. Sample injection system.   Sample injection into the chromatograph is normally made with glass syringes of varying sizes, but now gas sampling valves are also used. It is sometimes necessary to preconcentrate samples either by freeze-out or solid/liquid adsorption techniques so that the detector has sufficient sample material to respond much above its lowest detectable limit and hence perform accurate analyses.

Withdrawal of the sample gas from the source continuously and at a relatively high flow rate is practised often in order to minimize the time for possible interactions or reactions of the sulfur compounds with materials of construction. From this flow, a secondary stream of a smaller flow rate is usually selected, preconditioned, and then led to an automatic sampling valve of the chromatograph to be injected frequently for rapid analyses.

b. Column technology.   It is important to choose the proper column and packing (if used) in order to obtain successful analysis by gas chromatography.

Important variables in column technology are the column length and diameter, the solid support, the stationary or liquid phase, and the tubing material. Selection of a proper column is necessary in order to efficiently separate gaseous components of interest, because different gases have specific affinities for different packing materials. The liquid phase of a column separates the various components by differences in vapor pressure or polarity; in either case, molecules of greater molecular weight tend to remain in the column for longer periods. The degree of separation between components that can be achieved with a column increases with length and decreasing diameter of the separating column. For sulfur gas analyses with GC, the column materials that are sufficiently inert and independent of temperature effects (Adams and Koppe, 1967) include Teflon, 316 stainless steel, and glass.

For general use, solid phase support materials must be of sufficient inertness, porosity, uniformity, strength and must be easy to pack. The separation efficiency of the solid support is directly proportional to its porosity and surface area, but inversely proportional to its inertness. One of the more successful sulfur-separating columns is made of acetone-washed Porapak QS polymer and is described by de Souza et al. (1975b). This column, marketed as "Supelpak-S" by Supelco, Inc.,* is unaffected by water vapor, employs no liquid phase, and hence demonstrates no baseline drift even with temperature programming.

  c. Chromatographic detectors. Some of the chromatographic detection systems in use for sulfur gas analyses are thermal conductivity, microcoulometry, photoionization, flame ionization and flame photometry. The detectors used must have accuracy, stability, sensitivity, selectivity, durability, rapid response, minimum maintenance, and freedom from interfering substances. Thermal conductivity detectors are based upon the fact that different gases have different thermal conductivities. Two matched and heated metal elements are installed in two gas channels, one of which carries a reference gas, the other the sample gas. The difference in the cooling rates of the preheated metal elements by the reference and sample gas is a measure of the concentration of the sample gas. These detectors are not specific or sensitive enough for most pulp mill applications. Microcoulometric detectors are identical to the coulometric detectors described in Section II,E,4,a. However, they have smaller sample handling capacities and are generally unsuitable for source monitoring because they can be easily overloaded at high concentrations and could require frequent maintenance. The photoionization detector is suitable for those compounds that ionize by absorbing ultraviolet (UV) light. A chamber close to the UV source contains a pair of electrodes. When a potential is applied between the electrodes, the created electrical field drives any ions formed from the sample gas (by absorption of UV

---

*Supelco, Inc., Supelco Park, Bellefonte, Pennsylvania 16823.

light) to the collector electrode, thus generating a current that is proportional to the concentration of the ionized material. The detector is nonspecific, good for some organic compounds, and has a wide linear range.

The detectors most commonly used in the pulp and paper industry are the flame ionization and flame photometric units. Both types require hydrogen flames but are free from water interference and have excellent stability characteristics. The flame ionization detector is suitable for organic sulfur and non-sulfur-containing compounds, which can be ionized in flames. It is not sensitive to inorganic H_2S or SO_2. The most popular and useful detector is the flame photometric that comes either with a single or a dual flame (Section II,E,1) and is suitable for the determination of most sulfur and phosphorous compounds.

d. Calibration procedures. Calibration of gaseous monitoring instruments at frequent intervals is necessary in order to maintain their continued accuracy. A typical calibration procedure checks the response of an instrument to known concentrations of standard or span gases. Methods in use for preparing known gas concentrations in order of increasing popularity include rotating syringes (Rossano and Cooper, 1963), motor-driven syringes, flexible fabric bags, lecture bottles, known cylinder mixtures (Duckworth *et al.,* 1963), and permeation tubes (O'Keefe and Ortman, 1966; Duncan and Tucker, 1970). The latter have to be checked either by recording the absolute weight loss at a constant temperature over a known period of time or by using a primary standard such as the colorimetric method of West and Gaeke (1956) for SO_2 after the sulfur gas (if other than SO_2) is converted to SO_2 in the presence of O_2 in a furnace maintained at about $825 \pm 25°C$.

Another important item in the GC makeup is the use of suitable data collecting, handling, and processing systems. These are discussed in more detail in Section IV, near the end of this chapter.

3. ULTRAVIOLET MONITORS

These analyzers are based upon the principle of ultraviolet radiant energy being adsorbed by the sample gas at a fixed wavelength in reference to another nonabsorbing wavelength, when radiation from a UV source is passed through a windowed cell containing the sample gas. Light transmitted through the cell (Du Pont Instruments; Saltzman and Williamson, 1971; Lang *et al.,* 1975), usually of a long path length, for higher sensitivity, strikes a semitransparent mirror and divides into a transmitted and a reflected beam. In the measuring channel, light passes through optical filters passing only the measuring wavelength, e.g., 289 nm, for SO_2. The difference in the two signals obtained from the reference and measuring channels is linear with the gas concentration.

Some monitors have the range of measuring up to 2000 ppm of a single component such as SO_2 or TRS (after conversion to SO_2). Organic compounds such as phenols, ethylene, pinene, etc., found in kraft furnace emissions, can cause some interferences. Also, where measurements are made after conversion of TRS to SO_2, some odorless sulfur gases if present in the sample can be wrongly included in the odoriferous sulfur gas measurements.

4. ELECTROCHEMICAL ANALYZERS

a. Coulometric titrators. Sulfur gas analyzers based on coulometry (ITT Barton Instrument Co., 1967; de Souza and Prahacs, 1980) measure the electrical current necessary to maintain a fixed low concentration of free halogen, e.g., bromine, in a halide (hydrobromic acid) solution. As the sample gas consumes the free bromine, more of it is generated from the halide solution in order to maintain the original concentration. The sample gas is passed through a gas-tight titration cell (Fig. 6) through a porous glass sintered head into a solution of hydrobromic acid. A control box is used to govern the amount of current applied to a free bromine-generating electrode, the current being directly proportional to the bromine-consuming sulfur compounds. The electronic control box and the recorder can be placed as far as 120 m from the titration cell, which is usually placed as close as possible to the source being monitored. The magnitude of the cell response is different for the various sulfur compounds. This type of sensor can be used for analysis of TRS in the concentration range of about 4 ppb to 2000 ppm.

One coulometric instrument widely used in the pulp and paper industry is the Barton Titrator. It is relatively simple, gives real-time analyses, and is comparatively inexpensive. However, as mentioned by de Souza and Prahacs (1980) and Thoen *et al.* (1968, 1969a), it suffers from certain interferences or other problems when operated in either one of two modes usually employed: Mode I (without oxidation furnace)

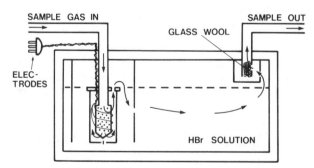

Fig. 6.   Cutout of a Barton titrator cell.

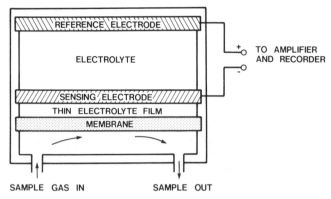

Fig. 7. Diagram of a simplified fuel cell detector.

(i)   Apart from reacting with sulfur compounds, bromine is also known to react with organics such as terpenes, phenols, ethylene, etc. (Altshuller and Sleva, 1962; Austin, 1971) present in many emissions, including those from kraft recovery furnaces. This results in positive errors that are in some proportion to the concentration of the reactive organic species present.

(ii)   The titrator responds differently to H_2S, SO_2, CH_3SH, $(CH_3)_2S$, and $(CH_3)_2S_2$. Since the relative proportions of these components are usually unknown (and variable), the "factor" used in the calculations, and thus the concentrations reported, can be in significant error.

Mode II (with an oxidation furnace used to convert TRS to SO_2 before measurement, in order to eliminate interferences from organics and to correct for the uncertainty of the species measured)

Any odorless sulfur compound found in kraft recovery furnace emissions after being oxidized to SO_2 would be wrongly included in the measurement of odoriferous sulfur emissions.

Also, in both modes of operation, if the sample flow rate through the titration cell "drifts," the reading will be in error; the magnitude of the error being proportional to the deviation from the flow rate used during the calibration run.

b. Fuel cell sensors.   The electrochemical transducer cell is a totally enclosed system (Fig. 7) equipped with two electrodes immersed in an electrolytic solution. Temperature and pressure are kept constant in the cell. The sample gas passes over a semipermeable plastic membrane through which the gas to be measured, e.g., SO_2, diffuses into the electrolytic solution. The diffused gas produces (Hann and Nylund, 1979) a change in the electrochemical potential across the sensing and reference electrodes that is directly proportional to the concentration of the gas.

Electrochemical membrane or fuel cell sensors are specific for either SO_2 or H_2S and can be used for monitoring reduced sulfur gas emissions. However, in the latter case, it is normally necessary to convert the reduced sulfur compounds to SO_2 in an oxidation furnace upstream of the detector. The membrane used in SO_2 measurement can be "poisoned" by H_2S or other reduced sulfur compounds and vice versa.

Electrochemical membrane cells specific for SO_2 are said to be relatively free from chemical interferences such as oxides of nitrogen, sulfur trioxide, water vapor, and organic compounds. They maintain a relatively stable calibration, without substantial drift in response for extended periods, with minimum maintenance. However, the electrolyte solutions in the detection cells become depleted with use, and cells must be replaced with new ones as the need arises. The major operating difficulties could be the presence of particulate matter, which tends to plug the membranes, water vapor condensation in the detector, causing erratic response, and the presence of H_2SO_4 mist, causing severe corrosion. The sensor is temperature- and pressure-dependent and could be subject to delayed response times.

F. Monitoring of Sulfur Dioxide

Detection systems used for continuous monitoring of SO_2 include electrolytic conductivity, electrochemical transducer conversion, coulometric titration, FPD, and ultraviolet spectrophotometry. The first four systems are all located outside the stack and require prior sample withdrawal, whereas ultraviolet spectrophotometry can be performed either internally or externally. Internal location of the detector eliminates any possible problems caused by air leakage or moisture condensation that might occur during sample withdrawal. External location is the only feasible approach for certain gas detectors that results in fewer problems with particulate interferences and high gas temperatures.

1. CONDUCTIVITY DETECTORS

Conductivity measurements are reported by Miller *et al.* (1968) for measuring total sulfur oxides (SO_2 and SO_3) in flue gas streams from digester blowpits and acid-making absorption towers in sulfite pulp mills. It is claimed that the system has been successfully operated over extended periods of time.

2. ELECTROCHEMICAL CELLS

A system used for continuous SO_2 monitoring with an electrochemical sensor has been described in the literature by Mathis (1973). It is reported that the system was successful during extensive field use. However, it is recommended that the modular transducer detection cells be replaced at intervals ranging from 6–18 months depending on cell design and SO_2 concentration level in the flue gas stream. See, also, Section II,E,4,b.

3. COULOMETRIC TITRATORS

Coulometric titration can be used for monitoring SO_2 emissions from coal- and oil-fired power boilers and sulfite pulp mill process sources where reduced sulfur compounds are not present. The measuring technique already described (Section II,E,4,a) is nonselective for SO_2 so that organic olefins and other materials if present in the sample stream can interfere. It is, therefore, necessary to add a combustion furnace upstream of the detector in order to oxidize the olefins from oil-fired boiler flue gases in order to prevent their interference. The method also requires removal of water upstream of the detector to avoid flooding the titration cell.

4. FLAME PHOTOMETRIC DETECTORS

This has been described in Section II,E,1 and will not be discussed further.

5. ULTRAVIOLET SPECTROPHOTOMETERS

Ultraviolet spectrophotometry is useful for the measurement of SO_2 concentrations either internally or externally of the stack duct. The method is relatively specific for SO_2 if an ultraviolet light source of the proper wavelength is used but is prone to interferences from water, organics, carbon dioxide, and particulate matter.

Internally located ultraviolet emission spectrometers for continuous monitoring of SO_2 in flue gases (Thoen *et al.*, 1969b) have been used. It is reported that the system has a minimum of interferences, is resistant to corrosion, and employs no moving parts. However, it is relatively insensitive to low SO_2 concentrations, displays sluggish response to rapid changes in concentration, and is prone to particulate buildup on cell windows.

Saltzman (1973) describes the use of an externally located dual beam ultraviolet spectrophotometric analyzer for continuous monitoring of SO_2 levels in flue gas streams from coal- and oil-fired power boilers and sulfite recovery furnaces. The detection system is a dual beam photometer in which ultraviolet light of 289 nm is passed through the sample cell to provide for specific SO_2 absorbance. Visible light, with a wavelength of 578 nm, is passed through a reference cell in order to minimize potential interference from NO_2. The system, heated to prevent water condensation, is found to be rugged and durable under field conditions and is said to maintain calibration on a stable basis for extended periods. To prevent small particles from being deposited on the detector cell surfaces, a glass wool filter is used in the sample line to remove the same.

Another available ambient monitor* is based on the principle that an SO_2 molecule emits specific light energy, which is measured with a photomultiplier tube, when the sample gas is submitted to a source of pulsed ultraviolet radiation. The monitor can be used to analyse source gases with suitable dilution of the sample. Some organic compounds are known to cause interference, but a "cutter" or filter can be used to reduce or eliminate this problem.

G. *Monitoring of Nitrogen Oxides*

Constituents other than TRS and SO_2 to be monitored in flue gas streams include NO, NO_2, and total oxides of nitrogen. These emissions from kraft and sulfite pulp mill process sources are not normally as significant as from power boilers, as explained in Section II. The major detection systems employed for continuous measurements of oxides of nitrogen, include electrochemical transducer membrane cells, ultraviolet spectrophotometry, infrared spectrophotometry, and chemiluminescence. Most of these systems are located external to the stack and therefore require sample conditioning devices.

1. ELECTROCHEMICAL CELLS

Electrochemical transducer membrane cell detectors are available for measuring either NO_2 or total oxides of nitrogen (NO plus NO_2) levels in flue gas streams. Details of these measuring devices have been covered in Section II,E,4,b. The sample gas is withdrawn with a sealed, leakproof stainless steel vacuum pump through a sample conditioner into an electrochemical detector (Mathis, 1973). The concentration of NO_2 and/or NO is taken to be proportional to the change in electrochemical potential across the cell with a readout of 0–10 mV. Readable concentration ranges are zero to 500, 1000, or 5000 ppm by volume.

2. ULTRAVIOLET SPECTROPHOTOMETERS

Ultraviolet spectrophotometry is useful for measuring oxides of nitrogen levels in flue gas streams. The detection principle is the same as for the SO_2 system described earlier (Section II,F,5) except that the wavelengths of the light beams

*Thermo Electron Corporation, Environmental Instruments Division, Hopkinton, Massachusetts 01748.

to the sample and reference cells are 436 and 578 nm, respectively, where NO_2 is the chemical compound being measured. A reactor at elevated temperature can also be used to convert NO to NO_2 in the presence of oxygen. The concentrations of NO_2 alone and NO_2 plus NO (after conversion) can be read sequentially in a timed cycle, the difference between the two readings being that due to NO.

3. INFRARED SPECTROPHOTOMETERS*

Infrared spectrophotometry can also be used to measure oxides of nitrogen. The principle of operation of a nondispersive infrared (IR) spectrometer (Jacquot and Houser, 1972);† is based upon the direction of infrared radiation through two separate and matched absorption cells, viz., a reference and a sample cell. The reference cell is sealed with a nonabsorbing gas such as nitrogen or argon. The sample cell receives a continuous stream of the gas to be analyzed. The amount of IR radiation absorbed by a particular gas component in the sample mixture, is proportional to the molecular concentration of that gas. The detector consists of a double chamber separated by an impermeable diaphragm. Radiant energy passing through the two absorption cells heats differentially the two portions of the detector chamber. The pressure difference causes the diaphragm to become distended and to vary a capacitance, which is measured electronically. The variation in capacitance, is proportional to the concentration of the gas components being measured.

There is not much field data available with the technique, which could be subject to all the potential interferences of water, organic compounds, etc., usually associated with IR methods.

4. CHEMILUMINESCENT ANALYZERS

Chemiluminescence is probably the best technique for measuring oxides of nitrogen. Stack gas is withdrawn from the duct (Heyman and Turner, 1976), conditioned to remove particulate matter and water vapor and then passed to a reaction chamber where NO is made to react with excess ozone (O_3) to produce light, the intensity of which is proportional to the inlet concentration of NO.

For the measurement of NO_2, the sample stream is diverted through a catalytic converter where NO_2 is reduced to NO. The latter, together with the original NO present in the sample gas, subsequently undergoes the chemiluminescent reaction with O_3. The resulting signal minus that of the original NO gives the concentration of NO_2.

---

*See also Chapter 10.

†Beckman Instruments Inc., Fullerton, California 92634; Mine Safety Appliances Co., Instrument Division, Pittsburgh, Pennsylvania 15235.

H. Monitoring of Oxygen, Carbon Monoxide, and Carbon Dioxide

1. OXYGEN

The principle of polarography is frequently utilised for the monitoring of O_2 in process streams. The sensor of the polarographic instrument contains a silver anode and a gold cathode, both protected from the sample by a thin membrane of Teflon. The membrane retains an aqueous potassium chloride (KCl) solution in the sensor, which serves as an electrolytic agent. Since Teflon is permeable to gases, O_2 diffuses from the sample gas to the cathode and takes part in an oxidation–reduction reaction:

at the anode:

$$4Ag + 4Cl^- \rightarrow 4AgCl + 4e$$

at the cathode:

$$O_2 + 2H_2O + 4e \rightarrow 4OH^-$$

Oxygen in the sample is reduced at the cathode when a potential is applied across the electrodes causing a current to flow, the magnitude of which is proportional to the concentration of O_2 in the sample gas.

Instruments based on the paramagnetic properties of O_2 are also available. These instruments are based upon the fact that oxygen is a unique gas in that it is attracted into a magnetic field (paramagnetic), whereas most other gases are slightly repelled out of a magnetic field (diamagnetic). Thus, by measuring the magnetic susceptibility of a sample gas mixture,* its oxygen content can be determined accurately.

2. CARBON MONOXIDE

One of the instruments used to monitor carbon monoxide (CO) is also based upon the principle of polarography. The sensor cell is a sealed unit containing a sulfuric acid electrolyte. Carbon monoxide gets electrically oxidized to CO_2 in the cell, the extent of conversion depending upon the partial pressure of CO in the sample chamber. The resulting electrical signal generated is compensated for temperature, amplified, and recorded.

Application of this monitor† is principally for ambient measurements, typically for such areas as garages, furnaces, and the like. But as in all such cases, it can be used for source monitoring after adequate dilution of the sample. Limitations of the monitor are the following:

*See the second footnote on p. 259.
†See the second footnote on p. 259; Econics Corporation, Sunnyvale, California 94086.

(i)   changes in atmospheric pressure will cause calibration to change;

(ii)   the sensor is gas-flowrate sensitive;

(iii)   high interference is caused by acetylene, ethylene, ethyl alcohol, nitric oxide and propane; however, it can be minimized by the use of an internal filter.

Monitors based on IR spectrophotometry* are also available.

Monitoring CO is practiced by some pulp and paper mills (Chamberlain *et al.*, 1977; Anson *et al.*, 1972)—in place of monitoring O_2—as a better indicator of complete combustion in a furnace. For this purpose a continuous Bailey Combustible Monitor is used. It is reported that there is a good direct correlation between CO and TRS emissions.

3.   CARBON DIOXIDE

Infrared spectrophotometry is widely practised for the continuous monitoring of CO_2. Basically, infrared absorption by the component of interest in the multicomponent sample is measured with spectrometers. The principle of operation is covered in Section II,G,3. Analysis of CO_2 in the parts-per-million to percentage concentration range is claimed* with such an analyzer.

III.   Monitoring of Particulates

Determination of particulate mass concentrations from stack flue gases is necessary if one is to meet current air pollution regulations, with increasing emphasis being placed on particle size distribution. Batch techniques are specified for determining particulate mass concentrations, emission rates, and size distributions. Continuous monitoring of particulate emissions is becoming prevalent as more accurate and reliable methods (Cooper, 1973) are developed.

A.   *Preliminary Considerations*

Particulate matter is normally defined as any material emitted into the atmosphere in either a solid or liquid state, including dusts, fumes, smoke, flyash, soot, tars, droplets, and mists. Changes in physical state with temperature can cause confusion in defining particulate matter for materials such as organic vapors and acid mists. Changes in chemical form, such as oxidation of SO_2 to H_2SO_4 in sampling train impingers after collection, can cause serious difficulty in interpreting the results obtained.

Particulate matter must be defined for either stack or standard conditions. The present definition accepted by the U.S. Environmental Protection Agency is that

*See the footnotes on p. 259.

particulate matter is any material that is collected on a filter of 0.45-μm porosity that has been heated to 121°C. Another definition for particulate matter specified by local agencies is material collected in liquid impingers at standard conditions.

An important prerequisite in particulate monitoring is that the sampling port must be located such that a representative sample is obtained. This can be reasonably achieved if the sampling port is located 8 diameters from any bends, constrictions, enlargements or obstructions in the flue duct. A second prerequisite is that the sampling rate be isokinetic, i.e., identical gas velocities must exist in both the flue gas duct and the sampling probe.

Sources of significant particulate emissions from the kraft chemical recovery system are the recovery furnace, the smelt dissolving tank, and the lime kiln. The recovery furnace is the largest potential particulate emission source. The major chemical constituent in the recovery boiler particulate emissions is Na_2SO_4, with smaller quantities of Na_2CO_3 and $NaCl$. The smelt dissolving tank vents and lime kiln exhaust gases are also sources of varying quantities of particulate matter consisting primarily of carbonate, hydroxide, sulfate, and chloride salts of calcium and sodium. Particle sizes from these sources can range from 0.1 to greater than 100 μm in diameter for uncontrolled emissions and from 0.1–10 μm where these sources are equipped with high-efficiency particulate control devices.

The two major types of particulate matter control devices employed for recovery furnaces are electrostatic precipitators and a wet absorbing type of scrubbers. The former are usually placed downstream of cyclone or cascade direct contact evaporators. Low pressure drop secondary wet scrubbers have been employed to supplement older and less efficient primary particulate collection devices at several existing mills to alleviate particle fallout in surrounding areas. Packed tower or showered mesh demister scrubbers are employed for particulate control on smelt dissolving tank exhaust gases, and venturi or cyclonic scrubbers are normally used for particulate control on lime kiln or calciner exhaust gases. The amount of particulate matter emitted from pulp mill process sources depends both on the process operating conditions and on the types and collection efficiencies of the control devices employed.

B. Batch Sampling

Batch particulate sampling is used to determine total particulate concentrations and emission rates from flue gas streams. The method with which the absolute mass of the collected solids is measured, is necessary for calibration of continuous monitoring devices, which are indirect indicators of the amount of particulates emitted. Batch sampling provides an average value for particulate concentration in a duct over a given time period but does not provide real-time instantaneous values. Collection methods for batch particulate sampling include filtration and liquid impingement (American Society of Mechanical Engineers, 1957; Devorkin *et al.*, 1972; Code of U.S. Federal Regulations, 1971). This method, although cumbersome, is widely used and is also a reference method advocated by the U.S. Environmental Protection Agency.

The primary collection stage consists of a glass cyclone (Fig. 8) and a 5-cm diameter glass fiber filter of 0.30–0.45-μm porosity that is heated to 121°C to prevent moisture condensation in the filter. However, condensation is allowed to take place in the secondary collection stage, which consists of a series of Greenburg–Smith water-containing impingers immersed in ice water. Condensation prevents flooding of the vacuum pump and gas meter placed downstream and helps absorb some potentially corrosive gases. In addition, the water vapor content of the stack gas is determined from the amount of condensate collected in the impingers. An S-type pitot tube located in parallel to the sample probe facilitates the maintenance of isokinetic sampling conditions. The system provides for efficient particulate collection, but the filters are subject to rapid plugging at high particulate loadings, resulting in large pressure drops across the filters and changes in the sample gas flowrates.

C. Size Distribution

Particulate size distribution in flue gas streams can vary from less than 0.1 to greater than 100 μm, with densities from 0.8 to 2.5 g/cm^3. Particle size distribution can be determined with some difficulty due to high humidity, water droplets, and the tendency of the particulates to coalesce or agglomerate. The most commonly used methods to determine particle size for emissions from pulp and paper mill sources are multistage cascade impaction (Bosch *et al.*, 1971) and membrane filtration (Walker, 1963).

D. Continuous Monitoring

Continuous monitoring of particulates from a source such as a recovery boiler is important in optimizing its operation. This could lead to reduction not only in particulate emission, capital, and operating costs of any abatement equipment used, but it could also help conserve many chemicals that would normally be lost through the stack. A few continuous particulate monitoring techniques practised today are well documented in the form of literature surveys (Cooper, 1973; Mitchell and Engdahl, 1963; Nader, 1975; Blosser *et al.*, 1974; Sem *et al.*, 1971a). Some continuous particulate monitors in current use are based on optical measurements (Wostradowski, 1977; Larssen *et al.*, 1972), beta-radiation attenuation (Sem *et al.*, 1971a; Barton and Turner, 1977), and charge transfer

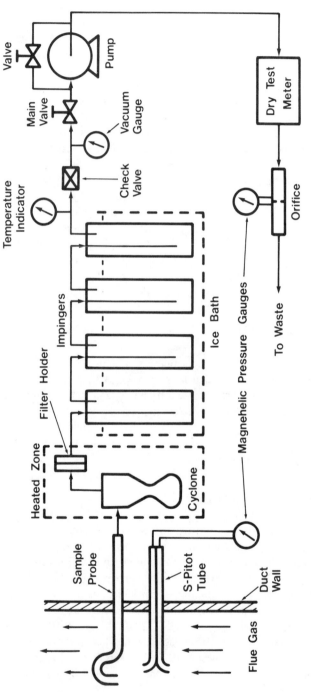

Fig. 8. Diagram of a sampling train used for collecting and measuring particulates.

(Azarniouch and Prahacs, 1977). Some other techniques that are being tried out include conductivity and specific ion determinations (Leonard, 1968; Tretter, 1969), piezoelectric crystallography (Sem *et al.*, 1971a),* optical lasers, lidar, and radar (Sem *et al.*, 1971b).

1. OPTICAL MEASUREMENTS

Optical devices in use for particulate monitoring include internally located bolometers (Gansler, 1968) and transmissometers (Larssen *et al.*, 1972; Beutner, 1973), where the degree of light attenuation is a function of particulate concentration in the flue gas duct. The system directs a light beam across the duct to a detector (single beam) or to a reflector and back to a detector (double beam) where the resulting electrical signal is amplified, transmitted, and recorded on a recorder. The sensitivity and accuracy of optical stack monitoring devices (Larssen *et al.*, 1972) are affected by the intensity and wavelength of light, path length, moisture content as a function of the temperature of the stack gas, particle size, color, and particle mass loading of the flue gas stream.

Optical particulate measurement devices do not measure particulate mass concentration and have serious limitations. However, they are simple, inexpensive, and relatively easy to operate and maintain; hence they are particularly useful for providing warnings of possible malfunctions of control equipment and nonconformance with stack opacity standards. The lenses of the detectors tend to be coated with fine dust with time, requiring frequent or continuous cleaning. Severe vibrations can lead to unstable readings. The devices are probably most suitable for stacks with low particulate concentrations following high-efficiency control devices where the particle size distribution and other physical properties are relatively uniform.

2. BETA-RADIATION ATTENUATION

In this technique, a beam of beta particles (electrons) is pased through a medium such as flue gas, resulting in some absorption and some reflection leading to a loss in beam intensity. Such a reduction is known as beta-radiation attenuation and is a measure of the mass of material through which the beam passes. The correlation of beta attenuation with mass is a function of the ratio of the number of electrons to nuclear mass per molecule. This ratio is between 0.4–0.5 for all elements except hydrogen (Sem *et al.*, 1971a; Larssen *et al.*, 1972); hence the correlation of beta attenuation with mass is virtually independent of composition of the particle material.

Solid material is collected on a filter, for a fixed time; then the filter with the

*Thermo Systems Inc., St. Paul, Minnesota 55113.

solids is placed between a beta source and a conventional Geiger detector tube (Barton and Turner, 1977)* The reduction in intensity of a beta-radiation beam passed through a clean filter alone and then through the filter loaded with particulates is measured to obtain by correlation, the mass of material collected over the filter for the duration of the sampling period.

Although these devices have good potential for use as mass monitors, they need to be improved in the fields of sample conditioning, sampling rates, transport mechanism for the particle-loaded filter tapes and sample transfer from the duct to the sensor without significant sample loss, for accurate determinations of particle mass loadings in the sample flue gas.

3. CHARGE TRANSFER

In this measurement,† a transfer of electrical charge from a particle in the sample stream to a sensing electrode takes place when they come into physical contact with each other. The charge transfer is then measured as a flow of electrical current from the sensor, with appropriate electronic circuitry.

This is a single-point measuring device (Azarniouch and Prahacs, 1977), with many potential problems. The probability and nature of physical contact between the particulates and the sensor depend upon the composition and surface properties of the particulates, the sensor, the flow characteristics of the sample stream and contamination of the sensing element. Also, the sensor may respond disproportionately to submicron particles that do not contribute proportionally to mass; it may be affected by the saturated or unsaturated nature of the sample gas and may be subject to invalid calibrations owing to changes in particulate shape and size distribution.

IV. Data Recording, Processing, and Printing

Most continuous analyzers are equipped with some form of data collecting device, whether they are in the form of chart recorders, data loggers, or computer systems. It is very important that the right choice be made in acquiring data reducing and reporting systems for any monitoring operation. This should be done with the intention of obtaining the final printout with the least amount of manual effort in the form desired, e.g., converting concentration data into practical units in terms of process rate such as kilogram/ton.

*Lear Siegler Inc., Englewood, Colorado 80110; Nucleonic Deta Systems, Irvine, California 92664; Research Appliance Company, Allison Park, Pennsylvania 15101.

†IKOR Inc., Environmental Technology, Burlington, Massachusetts.

The choice of a recording system for continuous emission monitors is governed by the measuring requirement or the completion of one cycle of operation (sampling and analysis). Data values taken instantaneously or obtained by integrating or averaging a number of data points over a fixed time period can be used to satisfy these measuring requirements. Consideration of both measuring and recording requirements will often dictate the choice of the total monitoring system.

Continuous monitors produce continuous traces on strip charts, data that is usually much more than is actually required, since regulations generally specify only the minimum number of data points to be averaged and recorded for a set time period.

Analog records, are often obtained with chart recorders in most source monitoring applications. They can be found either as strip or circular-chart recorders. The former offer great versatility, the latter are limited by the chart length and poor time resolution for measuring parameters at small values. The choice of a recorder depends upon a number of factors, some of which could be the type of unit, pen, paper take-up, the range of signal input, chart speed, accuracy, and response time.

Digital recorders or data loggers select and record intermittently either instantaneous or integrated values for a given time period. They do not process the data but merely record it at specified time intervals. They can be operated to sound an alarm or print out data once a set value is reached. Also, they can be modified to provide values that can be read directly, e.g., in parts per million or as percentages, and they can be interfaced easily with a computer for data processing. However, digital recorders are more complex and difficult to troubleshoot since the data are averaged and not instantaneous, as with analog recorders. Also, the difficulty of detecting trends with data loggers can be overcome by recording the digital data on cassette tapes, which can then be read off on a computer and graphically plotted for the time period of interest.

Data processors on the other hand can average and compute data rapidly in terms of emission standards, thereby eliminating the manual effort of data reducing. Generally, there are two data processing methods. The first interfaces the analyzer with an analog-to-digital converter which in turn is connected to the mill computer or data processing system. The latter accepts the digital signals and performs the necessary calculations for the final printout, an example of which can be seen in Table II. However, the plant computer may not have enough storage or programming capacity to handle the continuous monitoring needs. Also, the monitoring data could be either lost or difficult to retrieve should the computer malfunction or be rendered inoperable. The second method is usually more expensive since it is a dedicated system to process only the continuous monitoring data. These versatile systems can generally compute emissions, provide a summary report of the data for specified time periods, display the results at

TABLE II

An Example of a Report Printout

Date:
Source sampled:
Name of Analyst:

Operating conditions of gas chromatograph

Detector used:
Separating column used:
Volume of sample injected (ml):

Gas used:	O_2	H_2	Air	He (carrier)
Pressure (psig):				
Flowrate (ml/min):				

Temperature of injection port (°C):
Temperature of detector (°C):
Temperature of programming of oven, if practiced:

Sample No.	Actual Time	Dry concentrations (ppm)					TRS mass emission rate (kg/hr)	Indicate if TRS emission exceeds regulations
		H_2S	CH_3SH	$(CH_3)_2S$	$(CH_3)_2S_2$	TRS		

Atmospheric pressure (kPa):
Average stack gas pressure (kPa):
Average stack gas temperature (°C):
Average stack gas velocity (m/sec):
Averge moisture content of sample gas (%):
Average molecular weight of stack gas:
Average calculated volumetric stack gas flowrate (m^3/hr):

remote places, generate an alarm signal, and record the data either on paper or on magnetic tape.

Printing the final report on emissions can be set to desired formats often involving provisions to indicate emission rates in process terms and excess emissions that are over the amounts or guidelines set by the regulatory bodies for a particular pollutant. The report is usually made to the regulatory bodies with explanation/s for excess emissions. Operations or events such as soot blowing malfunctions, startups, shutdowns, blowbacks, etc., can be used in the explanations. Mention must also be made of the fact that there were no excess emissions during the reporting period, if such was the case.

V. Concluding Remarks

The pulp and paper industry has realized the importance of using automatic or continuous monitors in the curtailment of pollution of all types. In most instances, use is made of readily available continuous analyzers that are simple enough to be used as purchased. However, some need modifications in order to meet the requirements of a particular mill application. As time passes, the regulatory bodies are pressured into introducing more and more stringent standards for pollution control. This is primarily because a more informed public has been alerted to the dangers of the ever increasing load of pollution on our environment, as the race to industrialize steps up in all parts of the world. Automating the analyses of all types of process streams, including waste, can enable a processor to have a better control over his operations, thereby making it possible to cut back on wastage and pollution and, at the same time, improve the efficiency of the process. It is expected that by the turn of this century most processes and pollution monitoring systems will be fully automated.

References

Adams, D. F., and Koppe, R. K. (1967). *Environ. Sci. Technol.* **1**, 479.

Altshuller, A. P., and Sleva, S. F. (1962). *Anal. Chem.* **34**, 418.

American Society of Mechanical Engineers (1957). "Determining Dust Concentrations in a Gas Stream," Performance Test Code 27-1957. ASME, New York.

Anson, D., Clarke, W. H. N., Cunningham, A. T. S., and Topa, P. (1972). *Combustion* (March), 17.

Austin, R. R. (1971). *Tech. Assoc. Pulp Pap. Ind.* **54**, 977.

Azarniouch, M. K., and Prahacs, S. (1977). Evaluation of the IKOR in-stack continuous particulate emission monitor, *Can. Pulp Pap. Assoc., Tech. Sect., Environ. Improv. Conf.* November 1.

Barton, S. C., and Turner, E. N. (1977). A continuous particulate monitor for use in the pulp and paper industry, *Proc. Annu. Meet. Air Pollut. Control Assoc.*, *70th* June 20, Paper no. 77-34.2.

Beutner, H. P. (1973). Measurement of opacity and particulate emissions from stacks, *Miami Univ. Symp. Instrum. Continuous Monitoring Air Water Qual.* June 20.

Bhatia, S. P., Marentette, L. P., de Souza, T. L. C., Barclay, H., Wong, A., and Prahacs, S. (1975). *Pulp Paper Can.* **76** (C), T98.

Blomberg, L. (1976). *J. Gas Chromatogr.* **125**, 389.

Blosser, R. O., Cooper, H. B. H., and Megy, J. A. (1968). *Atmos. Qual. Improv. Tech. Bull.* No. 38.

Blosser, R. O., Kutyna, A. G., Schmall, R. A., Franklin, M. E., and Jain, A. K. (1974). *Tech. Assoc. Pulp Pap. Ind.* **57**, 74.

Bosch, J. C., Pilat, M. J., and Hrutfiord, B. F. (1971). *Tech. Asoc. Pulp Pap. Ind.* **54**, 1871.

Brody, S. S., and Chaney, J. E. (1966). *J. Gas Chromatogr.* **4**, 42.

Canfield, J. (1971). *Conf. Methods Air Pollut. Ind. Hyg. Stud.*, *12th* April 7.

Chamberlain, R. E., Lofkrantz, J. E., Norris, R. G., Smith, A. R., and Wostradowski, R. A. (1978). *Pulp Pap. Can.* 79 T44.

Code of U.S. Federal Regulations (1971). "Standards of Performance for New Stationary Sources," Part 60, Chapter 1, Title 40, Method 5. U.S. Environmental Protection Agency.

Cooper, H. B. H. (1973). The particulate problem: continuous particulate monitoring in the pulp and paper industry, *Miami Univ. Symp. Instrum. Continuous Monitoring Air Water Qual.* June 20.

Cooper, H. B. H., and Rossano, A. T. (1971). "Source Testing for Air Pollution Control." Environ. Sci. Serv. Corp., Wilton, Connecticut.

de Souza, T. L. C. (1982). "Truly Continuous Source Sampling Probe with Blowback for Gaseous Monitoring." Pulp and Paper Research Institute of Canada, Pointe Claire, Quebec, Canada (to be published).

de Souza, T. L. C., and Prahacs, S. (1980). *Pulp Paper* **54**, 124.

de Souza, T. L. C., Lane, D. C., and Bhatia, S. P. (1975a). *Pulp Pap. Can.* **76**, 73.

de Souza, T. L. C., Lane, D. C., and Bhatia, S. P. (1975b). *Anal. Chem.* **47**, 543.

de Souza, T. L. C., Wostradowski, R. A., Poole, R., Vadas, O., Bhatia, S. P., and Prahacs, S. (1978). *Pulp Pap. Can.* **79**, T242.

Devorkin, H., Chass, R. L., and Fudvrich, A. P. (1972). "Source Testing Manual." Air Pollut. Control District, Los Angeles, California.

Dragerwerk, H., and Drager, B. (1962). German Patent 1,113,918.

Duckworth, S., Levaggi, D., and Lim, J. (1963). *J. Air Pollut. Control Assoc.* **13**, 429.

Duncan, L., and Tucker, T. W. (1970). *Atmos. Qual. Improv. Tech. Bull.* No. 47.

Du Pont Instruments. Du Pont Model 464 source monitoring system for SO_2 and total reduced sulpher, Bull. No. 464. Wilmington, Delaware.

Ferguson, D. A., and Luke, L. A. (1979). *Chromatographia* **12**, 197.

Gangwal, S. K. and Wagoner, D. E. (1979). *J. Chromatogr. Sci.* **17**, 196.

Gansler, N. R. (1968). The use of a bolometer for continuous measurement of particulate losses from kraft recovery furnaces, *Annu. Meet. Air Pollut. Control Assoc.* Pac. Northwest Int. Sect., Nov. 22.

Hann, G. K., and Nylund, J. E. (1979). *Pulp Pap. Can.* **80**, T315.

Heyman, G. A., and Turner, G. S. (1976). Some considerations in determining oxides of nitrogen in stack gas by chemiluminescence analyzer, *Instrum. Symp.* May.

ITT Barton Instrument Co. (1967). "The Barton Model 286 Sulfur Titrator." Montery Park, California.

Jacquot, R. D., and Houser, E. A. (1972). *Adv. Instrum.* **27**, 730.

Lang, C. J., Saltzman, R. S., and DeHaas, G. G. (1975). *Tech. Assoc. Pulp. Pap. Ind.* **58**, 88.

Larssen, S., Ensor, D. S., and Pilat, M. J. (1972). *Tech. Assoc. Pulp Pap. Ind.* **55**, 88.

Leonard, J. S. (1968). *Atmos. Qual. Improv. Tech. Bull.* No. 35.

Maksimov, V. F., Bushmelav, V. A., Torf, A. I., and Lesohhin, V. B. (1965). *Bum. Prom-st.* **40**, 14.

Mathis, G. V. (1973). Application of an electrochemical cell to NO_x and SO_2 monitoring, *Miami Univ. Symp. Instrum. Continuous Monitoring Air Water Qual.* June 21.

Metallurgical Corp. (1975). "The Mott Inertial Filter." Farmington Industrial Park, Farmington, Connecticut.

Miller, A. M., Brown, J., and Abrama, R. (1968). Applied techniques of analyses for stack emissions, *Natl. Counc. Pap. Ind. Air Stream Improv. West Coast Reg. Meet., Oct. 2.*

Mitchell, R. I., and Engdahl, R. B. (1963). *J. Air Pollut. Control Assoc.* **13**, 11.

Nader, J. S. (1973). *J. Air Pollut. Control Assoc.* **23**, 587.

Nader, J. S. (1975). *J. Air Pollut. Control Assoc.* **25**, 814.

National Information Service (1976). "Environmental Pollution Control Pulp Paper Industry," Part I. Air. PB-261 708/2SL, p. 17-5. U.S. Dept. of Commerce, Washington, D.C.

O'Keefe, A. E., and Ortman, G. C. (1966). *Anal. Chem.* **38**, 760.

Rossano, A. T., and Cooper, H. B. H. (1963). *J. Air Pollut. Control Assoc.* **13**, 518.

Saltzman, R. S. (1973). Use of photometric analyses for ultraviolet analyzers for NO_x and SO_x, *Miami Univ. Symp. Instrum. Continuous Monitoring Air Water Qual.* June 21.

Saltzman, R. S., and Williamson, J. A. (1971). "Monitoring Stationary Source Emissions for Air Pollutants with Photometric Analyzer Systems." E. I. du Pont de Nemours & Co. Inc., Instrum. Div., Wilmington, Delaware.

Sem, G. J. Borgos, J. A., and Olin, J. G. (1971a). *Chem. Eng. Prog.* **67**, 83.

Sem, G. J., Borgos, J. A., Olin, J. G., Pilney, J. P., and Liu, B. Y. H. (1971b). State of the art: 1971 instrumentation for measurement of particulate emissions from combustion sources—Volumes I and II: Particulate mass, Reports APTD 0733 and 0734, Documents PB 202 665 and

PB 202 666. U.S. Environmental Protection Agency, Air Pollut. Control Office, Durham, North Carolina.

Thoen, G. N., and Nicholson, D. C. (1970). *Tech. Assoc. Pulp Pap. Ind.* **53,** 224.

Thoen, G. N., DeHass, G. G., and Austin, R. R. (1968). *Tech. Assoc. Pulp Pap. Ind.* **51,** 246.

Thoen, G. N., DeHass, G. G., and Austin, R. R. (1969a). *Tech. Assoc. Pulp Pap. Ind.* **52,** 1485.

Thoen, G. N., DeHass, G. G., and Baumgartel, F. A. (1969b). *Tech. Assoc. Pulp Pap. Ind.* **52,** 2304.

Tretter, V. J. (1969). *Tech. Assoc. Pulp Pap. Ind.* **52,** 2324.

Walker, C. G. (1963) *Atmos. Qual. Improv. Tech. Bull.* No. 19.

West, P. W., and Gaeke, G. C. (1956). *Anal. Chem.* **28,** 1816.

Wilby, F. V. (1969). *J. Air Pollut. Control Assoc.* **19,** 96.

Wostradowski, R. A. (1978). *Pulp Pap. Can.* **79,** T202.

9

Continuous Analysis of Oxygen in Coke Oven Gas

DAN P. MANKA

Pittsburgh, Pennsylvania

I. Introduction

The presence of oxygen in process streams can have a deleterious effect in various ways: polymerizable compounds are easily oxidized; entrained air increases the volume of the process gas that must be pumped through the system; and explosive mixtures can be formed. We shall consider the second and third of these cases, particularly with coke oven gas, where the entrained air increases the volume of gas to be processed, and a high oxygen content can form an explosive mixture with the 55–60% hydrogen normally present in the gas.

Coal is charged through three holes located on the top of a coke oven. As the coal touches the hot sides of the oven, large volumes of smoke form, which pour out of the charging holes into the atmosphere, thereby creating a pollution problem.

To alleviate this pollution, a steam jet is located in the goose neck leading from the oven into the collecting main. When coal is charged into the ovens, the steam, at approximately 100 psig, is turned on. This jet not only pulls the smoke into the oven and into the main gas stream but also pulls in large volumes of air. This air mixes with coke oven gas generated in the other ovens connected to the same collecting main. Since only a very small amount of oxygen is liberated in the coking of coal, every 1% of oxygen in the coke oven gas indicates that 5% of the total gas is entrained air. Thus, in a coke plant producing 100,000,000 ft³ of gas per day, 5,000,000 ft³ is air, which must be pumped through the chemical recovery system without producing any benefit. On many occasions the oxygen concentration level reaches 3–4% or 15,000,000–20,000,000 ft³ of the total gas is air.

The oven is charged through three openings located on top of the oven. The oven is filled to a height of 12 ft with 16.2 tons of minus one-eighth in coal. A leveling bar on the pusher machine levels the coal in the oven. As noted previously, all doors and openings on the oven are closed so that the coal is heated in the absence of air, otherwise the coal would only burn and would not form coke. Oxygen analysis of the gas exiting from the oven during the 16–17 hr coking cycle is 0.1% or less. A total of 6950 tons of coal is charged per day, forming 4600 tons of coke per day, and 73,300,000 ft³ of coke oven gas per day.

II. Coke Oven Gas Flow Diagram

The flow diagram of gas is practically the same in all coke plants. As more desulfurization processes are installed, these will vary in location depending on the type of system installed.

In the flow diagram in Fig. 1, coke oven gas rises from the coke oven a, through standpipe b, to gooseneck c, where it is brought into contact with flushing liquor (ammonia liquor). Tar and moisture are condensed. Ammonium chloride, a portion of the free ammonia, fixed gases, hydrogen cyanide, and hydrogen sulfide are dissolved by the liquor. The gas, liquor, and tar enter the gas collecting main d, which is connected to all the ovens of a battery. In some cases there may be two gas collecting mains to a battery. The gas, liquor, and tar collected from all the batteries are separated in tar decanters. The tar f separated from the liquor e flows to tar storage g. A portion of the liquor e is pumped to the gooseneck c on the top of each oven. The remainder of the liquor is pumped to the ammonia liquor still h, where it is brought into contact with live steam to drive off free ammonia, fixed gases, hydrogen cyanide, and hydrogen sulfide. As the liquor flows from the free still h to the fixed still i, lime or sodium hydroxide k is introduced to liberate free ammonia from ammonium chloride. Live steam, admitted at j, flows up through the fixed and free stills, and the ammonia, fixed gases, hydrogen cyanide, and hydrogen sulfide are added through l to the main coke oven gas stream ahead of the ammonia saturator r.

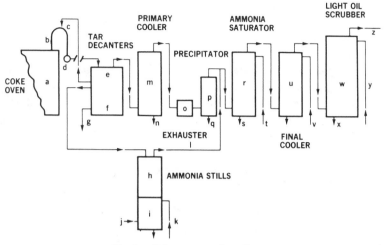

Fig. 1. Coke oven gas flow diagram.

The coke oven gas, separated from liquor and tar in e, is cooled indirectly with water in the primary coolers m. The fine tar that separates from the gas is pumped through n to the tar storage tank. The cooled gas is pumped by exhausters o to the electrostatic precipitators p, where additional fine tar is condensed and pumped through q to the tar storage tanks. The gas is contacted with a dilute solution of sulfuric acid in the ammonia saturator r to remove free ammonia. The ammonium sulfate laden acid flows through s to the ammonia crystallizer (not shown), where crystals of ammonium sulfate are separated, and the remaining sulfuric acid is pumped back to the ammonia saturator through t.

The ammonia-free gas flows to the final coolers u, where it is further cooled by direct contact with water. The water plus condensed naphthalene flows from the cooler through tar, which absorbs the naphthalene. The water is cooled and recirculated into the final cooler through v.

The cooled gas enters the wash oil scrubbers w, also known as benzole scrubbers, where it is brought into contact with wash oil, a petroleum oil pumped into the scrubbers through y. The aliphatic and aromatic compounds are extracted from the gas by the wash oil. The principal components are cyclopentadiene, benzene, toluene, xylenes, indene, and solvent, also known collectively as light oil. The benzolized wash oil is pumped through x to the wash oil still (not shown), where live steam strips out the light oil compounds. The debenzolized wash oil is cooled and returned to the wash oil scrubber. In some plants the light oil is further processed and fractionated into benzene, toluene, and xylenes and into a high-boiling solvent fraction. Naphthalene is also present in the light oil. Plants with low volumes of light oil do not have facilities for refining; therefore, the oil is sold to large refineries.

The gas from the wash oil scrubbers flows through z to a gas holder, which tends to equalize the pressure. Booster pumps distribute one-third of the gas for underfiring of the coke ovens and two-thirds to the steel plant where it is used as a fuel in the many furnaces.

As stated previously, a high oxygen concentration in the gas reacts with the unsaturated compounds to form gums that plug up orifices and valves. A concentration above 5% forms an explosive mixture with the 55–60% hydrogen normally present in the gas.

III. Sampling Location

The best location for monitoring the oxygen content of coke oven gas is after the booster pumps. At this point the gas is fairly clean, and it is at a pressure of at least 5 psig.

IV. Sampling

A 1-in. stainless steel pipe line is connected from the main gas stream after the boosters to a location ahead of the boosters. By this method, gas continuously flows from the 5-psig pressure to the 1–2-psig pressure line. Continuous sampling from this 1-in. pipe (Fig. 2) ensures a fresh sample for analysis. The gas is under sufficient pressure to provide a continuous flow; therefore, a pump is not required. Gas flows from the stainless steel pipe to a large water trap, where the major portion of the water entrained in the gas settles out. The drier gas flows from the top of the water trap to two scrubbers. Each of these are half full of a naphthalene absorbing oil, such as a petroleum oil or a light tar oil. After the gas has bubbled through the absorption oil in the two scrubbers, the gas flows to an empty scrubber to remove entrained oil. From this point the gas flows to a water-oil separator filter and finally through a fine filter to remove plus 5-μm particles. From the final filter, the gas flows to the large rotometer of the analyzer, Fig. 3.

V. Analyzer Operation

The oxygen analyzer is the type OA137 or OA269 manufactured by Taylor Instruments Co. (1976). This instrument measures the paramagnetic susceptibility of the sample gas by means of the proven, reliable measuring cell origi-

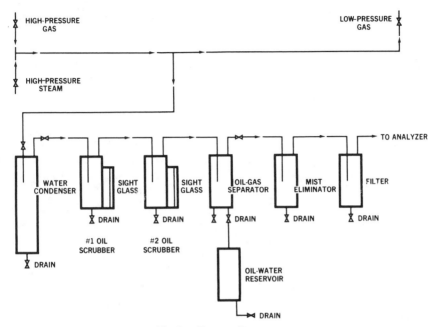

Fig. 2. Gas sampling system.

nally developed by BP Chemicals, Ltd. This cell consists of a nitrogen-filled glass dumbbell suspended on a platinum wire in a strong, symmetrical non-uniform magnetic field. Since it is normally slightly diamagnetic, it takes up a position away from the most intense part of the field. When the surrounding gas contains oxygen, which is paramagnetic, the dumbbell spheres are pushed further out of the field by the change in the field caused by the relatively strong

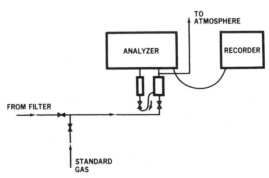

Fig. 3. Gas flow into analyzer.

paramagnetic oxygen. The image of a beam of light reflected from a mirror attached to the dumbbell falls on a differential photocell arrangement, which senses any displacement of the dumbbell. The strength of the torque is proportional to the paramagnetism of the surrounding gas, that is, the oxygen content. The signal from the photocells is amplified by an all solid state amplifier, which produces a proportional current output. This current flows through the single-turn feedback winding on the dumbbell to restore the dumbbell to its original position. Since this current is proportional and linear to the oxygen content of the gas sample, it is used to develop the millivolt output signals from the analyzer. This current feedback force balance design results in the outstanding accuracy and linearity of the instrument. The entire analyzer is located in an air pressurized box to keep dust out of the analyzer.

Because of the extremely linear relationship between the feedback current and the susceptibility of the sample gas, a proportional output voltage is developed, and various ranges can be obtained by means of a switched attenuator, namely 0–2.5%, 0–5.0%, 0–25.0%, and 0–100.0%

Linearity of scale also makes it possible to calibrate the instrument for all ranges by checking at only two points. For example, accurate calibration is obtained by using pure nitrogen for zero and air for setting the span at 21% oxygen.

Sample flow rate through the large rotometer in Fig. 3 is 30–40 ft^3 of coke oven gas per hour. A bypass flow meter is connected to the bottom of the large rotometer. This meter supplies sample gas to the analyzer. It is constructed so that the flow rate into the analyzer is maintained constant at 150 mliter/min over wide changes of supply pressure; therefore, a sample flow rate is maintained near the maximum to give the fastest response time.

Normally, variation in atmospheric pressure has little effect when the analysis is in the 0–5% range.

VI.   Interferences

The only common gases having comparable paramagnetic susceptibility are nitric oxide–nitrogen dioxide equilibrium mixture ($2NO_2$–N_2O_4) (NO_2 is the significant paramagnetic compound of this mixture) and chlorine dioxide (ClO_2). Chlorine dioxide is not found in coke oven gas, and the oxides of nitrogen are present in trace amounts so that they do not interfere with the analysis. If they are present in higher concentrations, 0.5% or more, the analyzer can be adjusted for the interference.

No other physical property of the gases has any significant effect.

VII. Daily Calibration

The sample gas is turned off and zero gas or nitrogen is turned on to flow directly into the analyzer. The nitrogen flow rate is adjusted to the flow used for sample gas. The attenuator is turned to 0 to 2.5%. When the analyzer has reached equilibrium in about 2 min, the needle on the meter is adjusted to zero with the mechanical zero screw. Dry air is used for span gas. The attenuator is set at 0 to 25%. Normal flow rates are maintained as used for sample gas and nitrogen. The analyzer comes to equilibrium in about 8 to 10 min. The span control should be adjusted to read 21% on the analyzer dial. When calibration is completed, the sample gas is turned on into the instrument and the flow rates are adjusted to the proper readings. The analyzer is calibrated once each day. During 35 days of continuous testing of coke oven gas for oxygen, calibration deviated by no more than 0.2% oxygen, indicating the high stability of the instrument.

The milliamp signal from the analyzer is fed to a Honeywell recorder equipped with a 24-hour round chart. This records the oxygen content as a peak each time coal is charged into the ovens.

VIII. Maintenance

As in all process stream analyzers, the sampling system requires cleaning periodically. The main problem in the coke oven gas line is accumulation of naphthalene.

The sequence of maintenance is identical each time. The sample gas is turned off at the sampling point after the boosters. The absorption oil is drained from the two scrubbers. The gas valve in the gas line after the final filter that feeds gas to the analyzer is turned off so that steam is not admitted into the analyzer. High-pressure steam is admitted into the main gas line near the sampling point. The valve on the bottom of the condenser is opened gradually so that naphthalene is steamed out of the line and through the condenser, which takes about 5 min.

Steam is slowly admitted into the absorption scrubbers and the oil separator and released to the atmosphere when the valves on the bottom of these scrubbers are opened. A steaming for about 5 min is sufficiently long to clean the system. The steam is turned off as are valves on the bottom of the condenser and scrubbers.

The scrubbers are half filled with fresh absorption oil. The gas valve in the line feeding the analyzer is opened, and the main gas valve is opened. The flow rate is adjusted on the rotometers. Normally the recorder is turned off during the calibration and steaming, because the span of the recorder is 0–5%. The 21% oxygen in the calibration air would tend to harm the recorder.

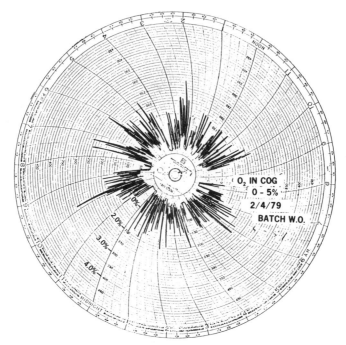

Fig. 4. Normal oxygen peaks.

IX. Results

The peaks in Fig. 4 show the oxygen content of coke oven gas each time coal is charged into the ovens. The oxygen range on the chart is 0 to 5%. The chart is a tell-tale of the sequence of oven charging, the time interval that steam is used in the ejectors. At times it even shows that steam was not turned off. Operation of coke ovens is poor when the ovens are not charged in sequence according to a preplanned timetable.

Irregular charging of the ovens causes variation in the flow of coke oven gas, variation in ammonia and tar contents, and variation in light oil content. Since the ammonia saturator and the light oil recovery operators do not realize these flow fluctuations, some ammonia and light oil is not recovered because the flow of sulfuric acid and absorption oil is not adjusted for these fluctuations.

Similarly, the hydrogen sulfide and hydrogen cyanide concentrations fluctuate so that the ability to recover these compounds for pollution control is affected.

Also, the operator of the reheating furnaces does not know how much air should be mixed with the gas for complete combustion.

In one instance, the oxygen content was over 5% for five hours during the night before it was discovered. The steam ejector was left on in an empty oven, which was being decarburized. All the air pulled in by the steam ejector was pumped into the main gas stream. Luckily, there was no spark to cause an explosion.

Charging of ovens with coal must be uniform and on schedule to produce a steady output of coke oven gas and to make the most coke per day. An examination of Fig. 4 shows gaps during the 24-hr cycle when ovens were empty and no coal was charged. This is particularly true during shift changes at 8 A.M., 4 P.M., and 12 midnight. There also are gaps for lunch at 12 noon, 8 P.M., and 4 A.M. When the oven operators are behind in their schedule, the ovens are charged rapidly and the oxygen content rises, because the operators are not careful to keep the steam jets on for the shortest possible time. This is shown very clearly in Fig. 5 when the operators were catching up between 3 A.M. and 4 A.M. These illustrations undeniably show the coke plant superintendent the reason for fluctuations in the production of coke oven gas and the reason his coke production is less than normal.

Figure 5 shows high oxygen peaks when the ovens were being charged very rapidly after a delay of several hours in the coal handling section of the plant.

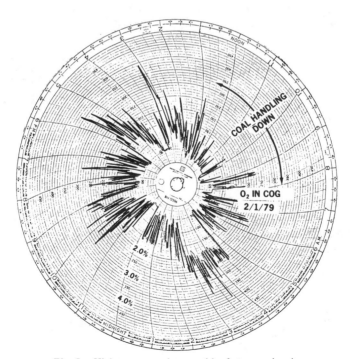

Fig. 5. High oxygen peaks caused by fast oven charging.

The operators charged the ovens so rapidly that they removed the lids from the top of the ovens and turned on the steam jets several minutes before charging.

These results are not from one plant but are typical of operation in most coke plants.

X. Relationship of Oxygen Content with Procedure for Coal Charging

There are three holes on top of each oven where coal is charged into the oven through chutes that are dropped into the oven from a Larry car above the ovens that holds the coal. In normal operation, the lids are removed from the holes, the steam is turned on, and the damper in the oven gas outlet is opened. The Larry car is positioned within 35 sec after the steam is turned on. The chutes are lowered, and the coal charging commences. This positioning takes another 20 sec. In this procedure, the oxygen content varies between 1.0 and 1.5%. However, a delay in positioning the Larry car, for example, 1½ min and charging commencing at 1 min, 50 sec, results in an oxygen content of 3.2%. The longer the delay from "steam on" to charging, the higher the oxygen content.

In another test, the Larry car is positioned above the lids on the oven holes. The steam is turned on; the lids are removed, and the chutes lowered. The oxygen content is 0.9% when charging begins within 20–25 sec after the steam is turned on.

In a third method, the oxygen is 1.0% or less for each charge. In this procedure all three lids are in place when the Larry car is spotted above the holes. Steam is turned on, the first and third lids are removed, and charging begun immediately. When charging is completed, the two lids are placed on the holes, and the second lid is removed and charging begun. When charging is completed, the lid is replaced on the hole. This method of charging is the best for keeping the oxygen content below 1.0%. The staff of each plant must determine which method is best for its operation. However, to be successful, the charging time from "steam on" must be limited to no more than 40 sec. A rigid adherence to this schedule by the plant personnel is necessary to maintain an oxygen level at or below 1%.

Reference

Taylor Instrument Company, Division of Sybron Corporation (1976). Industrial oxygen analyzer type OA137, File 18-4A, February 15.

10

Improving the Quality of Infrared Gas Analyzers

DAN P. MANKA

Pittsburgh, Pennsylvania

I. Introduction

There are several operations in the steel industry for which continuous stream analyzers are advantageous.

Customary control of chemical processes has been through analyses of snap gas samples by the Orsat method, a laborious and long procedure. The results are not available for a long time, and variations in gas composition cannot be determined to be of assistance to the operator. However, reliable and continuous infrared analyzers measure concentrations rapidly and can be used for computer control of the process.

Although infrared is the best method for analyzing these gases, the analyzers are subject to variation in (1) gas flow, (2) gas pressure, (3) gas temperature, (4) analyzer temperature, and (5) atmospheric pressure. These will be discussed in this chapter together with methods that were developed to reduce or even to eliminate these variations.

II. Sample Preparation

This subject is discussed in Chapter 11.

TABLE I

Effect of Gas Flow Rate on Thermal Conductivity and Infrared Analyzers[a]

H_2 (%)	CO (%)	CO_2 (%)	Gas flow rate (scfh)
2.71	24.15	14.41	3.15
2.71	24.02	14.36	2.85
2.72	24.2	14.43	3.25
2.72+	24.39	14.5	3.5 +
2.72	24.1	14.4	3.0
2.72	23.85	14.25	2.5
2.69+	23.59	14.15	2.0
2.69	23.25	14.02	1.5
2.69	23.05	13.97	1.25

[a] Constant pressure of 2 psig.

III.   Effect of Gas Flow

Table I illustrates the effect of flow on the CO and CO_2 concentrations when the flow rate is increased at a constant pressure of 2 psig. The effect of hydrogen is insignificant because the sensor is a diffusion type, which is not flow-sensitive. However, significant variation in concentration occurs with CO and CO_2. Thus, CO increases from 23.05%–24.15% when the flow rate is increased from 1.25 scfh to 3.15 scfh at 2 psig, and CO_2 increases from 13.97–14.41% at the same increase in flow rate.

IV.   Effect of Pressure

Pressure causes a similar result in the analysis. Variable pressure at a constant flow has little effect on the hydrogen detector. However, an increase in pressure causes CO and CO_2 to increase in concentration because there are more molecules per cubic foot of gas at higher pressure. These results are also caused by variations in atmospheric pressure.

The effects of flow and pressure are eliminated by a combination of regulators, as will be explained. Gas flowing through the analyzer system is increased to a predetermined pressure, which is above the maximum atmospheric pressure registered during the previous five years at the local weather bureau. This preset pressure is held within 1 mm by the special control system.

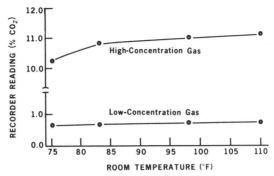

Fig. 1.   Effect of increased room temperature on CO_2 analyzer.

V.   Effect of Room Temperature

The instruments and the room must be temperature-controlled. The influence of a temperature variation, particularly on the infrared analyzers, is quite different, even for analyzers of the same type. In one analyzer, an increase in room temperature of 10°F causes a deviation of several tenths of one percent, whereas the effect may be negligible in another one of the same type. Figure 1 illustrates the increase in printout of CO_2 when the room temperature is increased from 75°F to 110°F, a normal increase when the analyzer is located in the plant in a room without air conditioning. The analysis of a gas containing 10.2% CO_2 increases to 11.0% and a gas containing 0.8% increases to 0.9% as the room temperature rises to 110°F. This deviation necessitates room temperature control to +2°F. Insulation and protection from draft is also important on connections where thermal effects occur. These are generally discovered by trial.

VI.   Control with Pressure Regulators

The best solution to the variations caused by atmospheric pressure, flow, and gas temperature is obtained by using pressure regulators. The gas sample is heated to 122°F in a coil located in a separate insulated box where the temperature is thermostated to ±0.1°F. The change in flow rate and atmospheric pressure is regulated by absolute pressure regulators located in the same heated box. The gas is preheated in a coil to 122°F and the pressure is reduced to 15 psig in the first 40-E-15* pressure regulator. Excess gas flows from the side of the regulator

*Moore Products, Cleveland, Ohio 44117.

to the atmosphere. The gas flows from the discharge end of the 40-E-15* valve to the inlet of a 43-20* absolute pressure regulator, where the gas pressure is reduced to 2 psig. Excess gas is vented to the atmosphere from the side of the valve. The pressure is constantly compared to the absolute vacuum section of the valve. This comparison maintains a constant pressure of 2 psig regardless of variation in the pressure of the incoming sample gas and to variations in atmospheric pressure. The sample then flows through a small rotameter set at 4 cfh into the H_2 cell, the CO_2 cell, and finally the CO cell. After analysis, the gas flows from the CO cell into the discharge end of the second 43-20* absolute regulator in the heated pressure control box.

Instrument air, at a pressure of 15 psig, is fed to a coil in the pressure control box where it is preheated to 122°F before the pressure is reduced to 2 psig in the second 40-E-15* pressure regulator. Excess air bleeds from the side of the valve to the atmosphere. The air leaves the discharge end of the valve and flows to the inlet of the second 43-20 absolute pressure regulator, where it is reduced to 1.5 psig. Air from the inlet of the valve and sample gas from the discharge end of this 43-20 absolute pressure regulator combine within the regulator and together bleed to the atmosphere through the side opening of the valve. Constant comparison of the sample gas pressure to the absolute vacuum built into the regulator maintains a constant flow and a constant gas pressure regardless of changes in atmospheric pressure. The maximum pressure maintained in the analyzer should be about 0.1 to 0.2 psig higher than the highest atmospheric pressure recorded at the local weather bureau during the past five years. In two such installations in process analyzers, the flow rate and pressure in the analyzer system have remained constant for several years using this system of control.

The analyzer room should be air-conditioned in order to maintain a constant temperature, and it should be pressurized with air to keep the room dust-free.

VII.   Interference by Fine Solids

In chemical processes, the off gases from a chemical reaction are analyzed continuously in order to determine the progress. An example of such a process is the blast furnace in the steel industry. Typically, the constituents measured are CO, CO_2, H_2, and occasionally CH_4.

It is usual to separate the gross solids from the top gas in an apparatus known as a dust catcher. The gas sample to be analyzed in the infrared analyzer is taken from a point downstream from the dust catcher, and the gas is filtered through successive filters designed to remove solid particles larger than 0.07 μm in size.

*See footnote on p. 285.

The gas so filtered is then passed through measuring units adapted so as to measure continuously and record the contents of the H_2, CO, CO_2, and CH_4.

Normally, a blast furnace operates under continuous blast and so continuously evolves gas. However, it is sometimes necessary to interrupt operations so that the evolution of gases ceases. When this happens, the gas analysis apparatus is cut off from the sample stream, usually automatically by shutting a valve in the sample pipeline, and the pumps are turned off. This allows atmospheric air to enter the analyzer through the exhaust pipe.

When the furnace operation resumes, it is standard procedure to check the calibration of the gas analysis apparatus before reconnecting it to the sample stream. This is done by passing a calibrating gas through the analyzers. Normally, the analyzer will give correct values in a period of about 2 min; it is then connected to the sample stream. Similarly, when the analyzer is checked against the calibrating gas in regular operation, it normally gives correct readings in about the same 2-min period.

It sometimes happens, however, that after an interruption of furnace operation, the gas analysis apparatus requires a period of time as long as one hour to come up to equilibrium with the standard gas. The analyzer will initially give CO and CO_2 content readings lower than the known contents of those compounds in the calibrating gas. The H_2 reading, on the other hand, will be above the H_2 content of the calibrating gas and will slowly reduce to the correct value. The reason for the higher H_2 content is the fact that CO and CO_2 have been adsorbed from the gas. At such times, the gas analysis system cannot be used to analyze the stream sample.

It is well known in the art of continuous gas analysis that in spite of the filters used to remove all solid particles from the gas stream being analyzed, a thin layer of fine dust builds up inside the gas analysis apparatus. Deposits of this sort from mill environments seem to be impossible to prevent. This dust appears to be a type of molecular sieve or activated silica. This compound adsorbs CO and CO_2 from a gas stream containing these constituents until it is saturated. If it is then exposed to a gas stream or an atmosphere that does not contain these constituents, or that contains moisture, the CO and CO_2 are desorbed. During normal analysis of dry sample gas in the analyzers in which the dust is deposited the adsorbent compound remains saturated with CO and CO_2, and the analyzer indicates the concentration correctly. If the analyzer is switched to a calibrating gas, it reads correctly in a period of 2 min. If, however, atmospheric air or other gas not containing CO and CO_2 gets into the analyzer, the activated silica or molecular sieve desorbs its CO and CO_2 to that gas. When a calibrating gas is reintroduced into the analyzers, these activated particles absorb these constituents from the calibrating gas until they are saturated. This accounts for the 1-hr delay in reaching equilibrium.

The amount of this fine dust in the analysis system is very small so that the entire analyzer system has to be cleaned to recover a sufficient quantity for analysis. This phenomenon appears to occur when the lining of brick and mortar in the blast furnace or in a heat treating furnace is nearing the end of its cycle, when the vessel must be relined.

The solution to this problem is to trap a gas containing CO and CO_2 in the analyzer during a shutdown so that the active particles are always at equilibrium with a gas containing CO and CO_2 (Manka, 1972). This solution necessitates a normally open solenoid valve in the analysis system immediately after the last infrared analyzer and in the sample line immediately ahead of the inlet of the calibrating gas. When sampling is discontinued, the two solenoid valves are closed, thereby trapping a gas containing CO and CO_2 in the analyzer. This solution has worked very well by maintaining the solids saturated with CO and CO_2 at all times.

Infrared is the best method for analyzing gases continuously if the absolute pressure regulators and solenoid valves are built into the system to eliminate fluctuations due to changes in atmospheric pressure and to reduce the interference by activated absorbing compounds. Each gas is analyzed continuously and easily compensated for interferences by other gases and moisture in the gas mixture. The manufacturer of the infrared analyzer must know the composition of the gas and moisture content to make proper mixtures in the infrared reference cell.

Reference

Manka, Dan P. (1972). U.S. Patent 3,673,854 assigned to Jones and Laughlin Steel Corp., Pittsburgh, Pennsylvania, July 4.

11

Waste Gas Analysis Techniques

DAN P. MANKA

Pittsburgh, Pennsylvania

I. Introduction

The customary control of chemical processes has been through operators' experience and through analyses of snap samples. Analyses of these samples usually require several minutes. When the results are finally available, considerable time has elapsed since the reaction occurred. In rapid reactions it is practically impossible to follow adequately the course of the process by these samples and to detect variations that significantly relate to control of the system.

In most reactions, the waste gas or off-gas composition relates closely to its progress. In a batch process, specific component composition in the off-gas is significantly low but rises rapidly as the reaction gathers momentum. It remains at a high plateau until chemical reactants in the media are depleted to lower levels, thereby decreasing the reaction rate. This decreasing rate is illustrated by decreasing composition of the pertinent components in the off-gas. Analyses of waste gas in the basic oxygen process in the steel industry is a typical illustration of a batch process. In a continuous system, such as the blast furnace in the same industry, concentrations of the significant components in the off-gas "ride" on the crest of a plateau indicating maximum reaction rate.

Instruments are available that analyze gas continuously or semicontinuously. Most reactions in steel processes are sufficiently rapid to require continuous analyzers. The best method for the continuous analysis of CO and CO_2 is to use infrared instruments: oxygen by the paramagnetic instrument, hydrogen by thermal conductivity, and moisture by the dewcel or by the thermoelectric unit.

Before we undertake a review of the many segments associated with the development of a reliable automatic gas analysis system, we shall consider three significant questions:

1. Will qualitative results from the installed system be adequate?
2. Will the system be upgraded to provide more quantitative results, which will be of assistance to the process operators?
3. Will the system be of high enough quality to present reliable and reproducible data, which are imperative for automatic process control?

If qualitative results are adequate, a simple system may be sufficient. Elimination of some causes of analyzer instabilities must be considered in a system that is to provide more quantitative results.

We shall discuss the third system, which is the most difficult—the one designed to provide reliable analyses as one critical segment of a process control. The actual control depends on the specific process, which is beyond the scope of this chapter.

A. *Specifications*

Before proceeding, it is well to establish the known specifications of aims of the finalized systems, namely:

1. Analytical accuracy and stability. The resultant analyses of small volumes of gas must be applied to the large volumes of gas in the process, which can be

300,000 ft³/min. Deviation of the analyzer by 0.2% causes serious problems in the final calculation. Therefore, specifications for maximum drift of the analyzers between calibrations is set at 0.15%. Accuracy must be based on the best available primary standards.

2. The filters must thoroughly clean the gas even at the high flow rates that are required to minimize response time. The filters must give reasonable service time before replacement is necessary.

3. The system should be compatible with plant operation. An example is downtime in the blast furnace when steam is injected into the dust catcher.

4. Signals should be provided throughout the system to alert the computer of malfunctions.

5. Automation permits computer control of various functions.

6. High availability and stability should be striven for.

II. Blast Furnace Process

The block diagram in Fig. 1 illustrates the sampling system in the blast furnace. Dirty gas is extracted from the duct and pulled through the filter by a pump. A small portion of the clean wet gas after the pump is fed to the moisture analyzer but the major portion is cooled in a refrigerator to remove the bulk of the moisture before it is pumped to the analyzers.

The relatively dry gas flows approximately 150 ft from the pump shed to the analyzer room. The schematic is shown in Fig. 2. By means of a bypass valve the second pump boosts the gas pressure sufficiently to force the gas through a final drier and into the analyzers. A major portion of the gas extracted from the plant duct is vented to the plant duct through the bypass valve. Only a small portion of the total volume is used for analysis. A large volume of gas is pumped to decrease residence time in the cleaning system because the analyzers are located at a distance of about 300 ft from the plant duct.

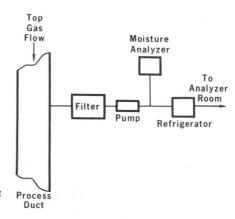

Fig. 1.   Systematic blast furnace sampling system.

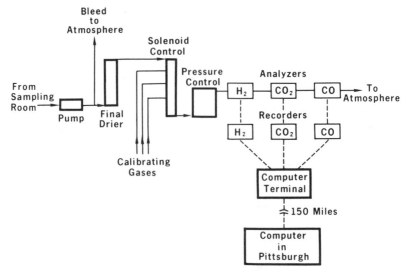

Fig. 2.   Schematic of an analyzer system.

The volume of gas apportioned for analysis is dried to a moisture concentration of a few parts per million. A solenoid valve system automatically or manually allows selection of blast furnace gas or any one of three calibrating gases to flow through the pressure controller and into the analyzers. All gas flows through the hydrogen cell, the CO_2 cell, and finally the CO cell before it is vented to the atmosphere. Signal lines are connected from each analyzer to the respective recorders. Each recorder is equipped with a transmitting slide wire, which furnishes a signal to the computer terminal. From the terminal the signals travel to the main computer located in the central computer building.

With few exceptions, a well-designed analyzer system meets the requirements of several processes. However, a sampling system is not predictable. Although the same raw material is used, the particulate matter in the gas from one furnace may vary significantly enough from the dust in a second furnace to necessitate redesign of the filtering media. The difference is not in the quantity of dust, but in its composition and physical characteristics.

A.   Sampling System

We shall now examine the significant sections of the sampling system.

1.   LOCATION OF THE SAMPLING PROBE

The sampling probe is usually located at a point in the waste gas duct system where the gas is well mixed but sufficiently close to the reaction vessel to provide a fast response time from the vessel to the analyzer. Louvers in the duct, elbows with built-in fins, and tees act as gas mixers. Because a reasonable response time is acceptable in the blast furnace system, the gas is sampled immediately after the dust catcher.

We have learned from experience that gas must be sampled continuously in order to obtain and maintain availability of analyses at 95% or more. Therefore, our policy is to install two sampling probes and filters to permit cleaning of one system while the other is sampling gas.

2. Filter System

Dust particles vary from one process to another and even from one reaction vessel to another. The basic filter system is decided upon based on a study of the particles from one reaction vessel. In general, this system is then applicable to all vessels of the same process, but variations do occur.

Dust particles in blast furnace gas are heterogeneous. Some are greasy; some are carbonaceous. The particles vary depending on the burden, the coke, and the limestone. When oil is also injected into the furnace, the filtering process becomes even more complicated.

Considerable testing and a knowledge of filter media aid immeasureably in the development of this portion of the system. Our experiments consisted of an actual filter installation on the blast furnace gas from the plant duct at a gas flow rate we considered necessary for the finalized system. Some filters were useless; others had a short life; and some could not be cleaned. The finalized system was based on the results of these tests.

3. Gas Temperature

A second important factor in the system is temperature. Moisture in the gas must never condense on the dust particles; if it does, a concrete-like solid is formed, which is removed with great difficulty. This is true for dust in the probes as well as for dust on the filters. Therefore, the temperature of a dirty gas line and filter must be maintained well above the expected maximum dew point of the gas.

4. Pump

The pressure of most process streams is insufficient to force gas through the filters and to maintain flow to the analyzers. A pump for extracting gas from the process duct is usually located on the clean gas side immediately after the filters. This pump must be oiless, greaseless, and leak-proof. Oil or grease from a pump interferes with the analyzers. It must be capable of pumping low gas flow rates and of operating at a high temperature in order to prevent moisture condensation.

B.   Gas Preparation

1.   MOISTURE REMOVAL

The sample gas requires further treatment before it can be allowed to flow to the analyzers. The bulk of the moisture in the gas is removed by refrigeration to a moisture content of 0.8% by volume. This is sufficient for an ordinary analysis, but for the precise system that we are discussing, moisture must be removed to the parts per million range so that absolute comparison can be made to the dry calibrating gases.

In a continuous process the final drier must be capable of reducing the moisture to the parts per million range without adsorption of CO and CO_2 from the gas. A regenerative drier is mandatory with no loss of gas analysis time during the regenerating cycle. This is done with Drierite in a regenerative system. While one section of Drierite is drying the sampling gas, the other section is being heated and purged with a portion of the dry gas to regenerate it.

C.   Stable Analyzers

1.   INFRARED ANALYZERS

Selection of suitable analyzers is another major decision in the development of a reliable system for process control. This is especially true when the initial drift specifications for each component are set at 0.1% or 0.15% at the maximum between consecutive calibrations.

The stability of infrared gas analyzers marketed by various manufacturers is affected by variations in (1) gas flow and atmospheric pressure, (2) gas temperature, and (3) room temperature. Methods to eliminate these variations are discussed in Chapter 10.

D.   Moisture Determination

Moisture in BF gas is analyzed in the clean but wet gas stream ahead of the refrigerator unit. This analysis is very important, particularly on cold blast air. The usual method of continuous determination consists of a Dewcel coated with a hygroscopic salt, such as lithium chloride, which is sensitive to changes in moisture content. The gold wire and glass tape section is coated with the salt. This cell, illustrated in Fig. 3, is sufficiently accurate for low concentrations but inadequate for high concentrations. Because some salt is continuously washed off the element, particularly at high moisture concentrations, the analysis suffers as salt concentration decreases. For this reason, the cell must be recoated often. A newer method does not depend upon a hygroscopic salt. It cools the sample

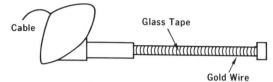

Fig. 3.   Dewcel for measuring moisture in blast furnace gas.

gas to the temperature at which moisture condenses from the gas. The temperature measured at the point of condensation is the dew point of the gas. This temperature corresponds to a specific moisture content.

E.   Operational Causes of Instability

We have discussed the major known causes of analyzer instability and methods to minimize or eliminate them. There are variations in external and internal operations that also affect the entire system.

1.   CONTROL OF FILTER SYSTEM

In the blast furnace system the filters will eventually become clogged. If the pump continues to operate in the absence of any gas because of a plugged filter or a malfunctioning solenoid valve, the pump will be damaged and must be replaced. To protect the pump and to signal a malfunction, a control system is built into the pump inlet line. When the vacuum in this line reaches a preset value, a controller automatically turns off the pump and signals the computer. The computer program switches to the second sampling line. If the second line also develops the same symptom in a specified time, the computer terminates all sampling by closing the sampling solenoid valves in the filter system. This action causes a vacuum to develop in the sample gas line, which automatically turns off the pump. The computer prints out a message stating that the filter system is clogged and gas sampling has been discontinued. This control system is an exceptionally good one. It continuously monitors the status of the filters; it prevents pump damage due to a plugged filter or to a malfunctioning sample solenoid valve; and it automates the sampling operations.

2.   STEAM IN DUST CATCHER

The blast furnace must go down for replacement of tuyeres and for other repairs. During this downtime, steam is continuously injected into the dust catcher and the ducts as a protective measure against the possibility of an explosion. If the sample system continues to extract gas during this operation, a large volume of moisture will flow through and condense on the filters. As previously stated, a mixture of condensed moisture and dust accumulated on the filter element forms a concrete-like mixture, which cannot be removed easily. Therefore, the sampling system must be protected from this steam. A control, connected to the plant steam line feeding the dust catcher, senses the pressure increase when steam is injected into the furnace ducts and turns off the sampling system. Simultaneously, portions of the analyzing system are turned off. When steaming of the dust catcher is terminated, the control senses the low pressure and activates relays to restart the gas sampling system. Gas is passed through the plant ducts for a period of about 10 min to flush out the moisture. Then the computer energizes the sampling system.

3.   EXCESS MOISTURE IN PREPARED GAS SAMPLE

The analyzers become useless when liquid water enters the analyzing chambers. The analyzing section must be taken apart for cleaning and, in some cases, certain parts must be replaced. Usually, the presence of liquid water in the sample gas is caused by a malfunction in the refrigerator whose function is to remove the bulk of the moisture in the gas. To detect the presence of excess moisture, a special probe or flood sensor is inserted in the gas line after the refrigerator.

When moisture condenses on the probe, solenoid valves are energized to divert the wet gas to the atmosphere outside the analyzer building, thereby preventing the access of moisture into the analyzer. The flood controller also generates a signal, which notifies the computer that the sample gas contains excessive moisture.

F.   Relay of Gas Analyses to the Computer

The computer must continuously know the composition of each gas in order to control the process. The analytical results for each component are fed to the computer through a transmitting slidewire connected to the corresponding recorder. The transmitting wire for the full range of the recorder is powered from zero to the maximum voltage stipulated for incoming signals to the computer.

In the blast furnace system the voltage from the transmitting slidewire is fed to an analog-to-digital converter for conversion to a transmission code. This code is transmitted through telephone wires to the computer located at the main computer building.

G.   Calibration

Although the major contributors to analyzer instability have been eliminated or minimized, there are a number of uncontrollable instabilities which necessitate a

periodic calibration of the analyzers. The time interval between calibrations must
be short enough to ensure close control of the instruments but long enough to
allow sufficient measuring time for the process gas.

In the blast furnace process, the computer initiates an automatic standardization every 8 hr. The analyzer curves for CO, CO_2, and H_2 are checked with
calibrating gases at the low, middle, and high concentrations that prevail in the
process gas.

The accuracy of the analyzers is dependent on the accuracy of the standardiz-
ing gases. Two types are used: calibration gases and primary standards. The
primary standards are prepared on a weight basis and analyzed by mass spectrometry. The accuracy of these gases is within 0.5% of the component concentration. The analyzers are initially standardized with the primary standards.
Concentrations in the daily calibrating gases are checked against the primary
gases in the infrared analyzers, and the resulting analyses are attached to each
cylinder. The analyzers are then routinely checked against these daily calibrating
gases. The accuracy of these daily cylinders is periodically compared in the
analyzers to the concentrations in the primary standards.

III. Basic Oxygen Process

We shall now briefly examine the basic oxygen process (BOP). The schematic
of the sampling and analyzing system is given in Fig. 4. Gas flows from the plant
duct through the filters to the pump. After the pump, a portion of the clean wet

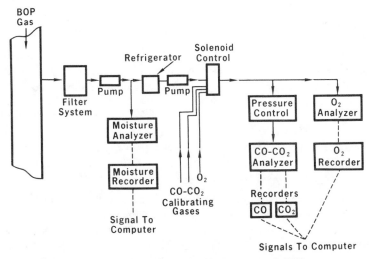

Fig. 4.   Schematic of sampling and analyzing systems in BOP process.

gas is diverted to the moisture analyzer. The major portion flows through a refrigerator to remove moisture to approximately 0.8% by volume. The gas pressure is boosted by a pump and diverted to the paramagnetic oxygen analyzer and to the CO_2 infrared analyzer. If CO is also present in the process gas, both CO and CO_2 cells are included in the same analyzer cabinet. The recorders are equipped with transmitting wires which furnish signals to the computer located approximately 500 ft from the analyzers.

A. Sampling System for Basic Oxygen Process

A response time of 20 sec is critical in the BOP process, therefore, sampling must be located in the dirty gas stream after the first thorough mixing.

The composition and physical characteristics of the dust particles are considerably different from those in the blast furnace. Selection of filter media is based on the dust characteristics. Again, we believe that two sampling probes and two filters are necessary in order to permit cleaning of one system while the second is sampling gas.

B. Stable Analyzers

1. INFRARED ANALYZERS

Discussion of the CO and CO_2 analyzers was covered previously in the blast furnace section, Section V, A and Chapter 10.

2. OXYGEN ANALYZER

Control of the BOP process requires knowledge of the oxygen concentration as well as those of CO and CO_2. Generally, oxygen is determined by the paramagnetic method. For accurate results this analyzer requires temperature and pressure control, as discussed in Chapter 10.

C. Relay of Basic Oxygen Process Gas Analyses to the Computer

The analytical results for each component are fed to the computer through a transmitting slidewire connected to the corresponding recorder. The voltage signal is fed from the recorder to the computer located at a distance of approximately 500 ft.

D. Calibration

The analyzers are calibrated manually or automatically. The CO and CO_2 analyzers are standardized at two points of the concentration curve, one in the neighborhood of the low end and the other in the neighborhood of maximum concentration in the process gas. The O_2 analyzer is also checked at the high and low end with calibrating gases.

IV. Performance of Basic Oxygen Process System

A. Sampling System

The BOP system has been operated for several years. In the installation the filters cleaned dirty gas from as many as 2000 heats at flow rates of 60–70 ft³/hr before replacement of the filters was required.

Plugged probes are very rare; they are usually caused by electrical failure.

B. Analyzing System

In the installation the special controls maintain a constant flow and pressure of gas in the analyzer. Readjustments have been infrequent since the original settings. Another reason for this fine control is the cleanliness of the sample gas which is thoroughly filtered before entering the analyzing system.

C. Stability

The constant flow and pressure, room temperature control, and clean gas have contributed to the stability of the analyzers. Daily calibration of the infrared analyzers for 30 consecutive days showed an average drift of 0.1% or less per day. Because the drift is low, plant operators calibrate every two days.

The average daily drift in the oxygen analyzers is approximately 0.2% per day.

V. Performance of the Blast Furnace System

A. Sampling System

The filtering media, suggested by the characteristics of the dust particles, has capably cleaned the gas even at flow rates of 120 to 140 ft³/hr.

The control of sampling when steam is injected into the dust catcher is operated automatically. This control has prevented countless failures of filters due to
condensation of excess moisture on the dust-laden element. Additional benefits
come from longer gas analysis and less manpower. Gas analysis is lost and
manpower is needed to replace water-damaged filters and to heat the new filter.
Analysis time is vital for control of a process.

Similar benefits result from the pump control. A pump cannot extract blast
furnace gas through a plugged filter, but it will be damaged. This control is the
safest and most logical signal available to the operator that filters need cleaning
or replacement.

The computer scans for signals from these controls, and it makes the necessary
change when a signal is detected.

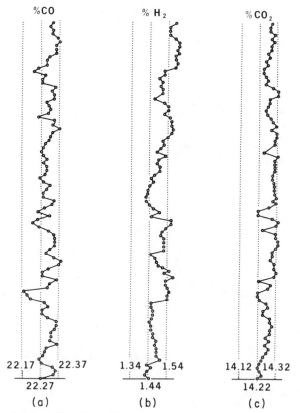

Fig. 5. Stability of (a) CO, (b) H_2, and (c) CO_2 analyzers in blast furnace process.

B. Analyzer System

A deviation of 0.15% by any one of the nine gases measured in a single calibration can void the analyzers.

Stability results are based on those obtained in a plant trial. After the initial setting of the recorders to the correct values at the beginning of the trial, the analyzers continued to operate without further adjustment. In 31 days of continuous operation, one gas in 94 calibrations was out of limits, deviating by 0.16%.

Stability of the three instruments during this test is best demonstrated in Fig. 5. This figure shows a plot of the actual drift of one gas in each analyzer between each 8-hr calibration. The total drift of approximately 0.2% in 31 days verifies that a stable system can be achieved.

Since the end of the trial, the analyzers have continued to operate at the same stability.

VI. Availability

A. Basic Oxygen Process System

When correct operation of the analyzers was developed, availability was more than 90%.

Availability of the sampling system is also high provided filters are replaced when they are almost plugged and general maintenance is continued.

B. Blast Furnace System

Availability of the analyzing system in the trial was 99%. Since the end of the trial, downtime has occurred mainly to check the calibrating gases against the primary standards.

Availability of the sampling system is also high. Downtime occurs for replacement and reheating of filters.

In conclusion, the development of a stable analyzing and sampling system is a difficult task, but it can be achieved. We believe that our systems are properly engineered for process control in the iron and steel industry. With modification, these are applicable to other processes.

12

Continuous On-Line Monitoring of Total Organic Carbon in Water and Wastewater

R. B. ROY, J. JANSEN, and A. CONETTA

Technicon Industrial Systems
Tarrytown, New York

I. Introduction

Determination of the concentration of organic carbon in natural waters, industrial waste effluents, sludges, and sediments is one of the most important of environmental analyses. Typically, these determinations are used to monitor organic pollutants in industrial wastewaters and industrial waste treatment processes. When the analysis is performed on homogenized samples, so that particulates and dissolved carbon from both organic and inorganic sources are determined, the measurement is called total carbon (TC). When samples are pretreated with acid to remove response from inorganic carbon, the measurement is called total organic carbon (TOC). When both particulates and inorganic carbon response are removed prior to sample analysis, the measurement is called dissolved organic carbon (DOC).

The measurement of the levels of total organic carbon indicate both the potential oxygen demand and the degree of pollution. The presence of organic carbon in water or other inorganic materials can seriously affect the proper functioning of these materials. For instance, organic compounds in lakes, rivers, and streams can deplete oxygen needed to support aquatic life. Most organic compounds are biodegradable, and in the process, use up the dissolved oxygen sources, resulting in undesirable changes in aquatic life forms. In most cases the effect due to oxygen depletion is directly proportional to the amount of organic materials present in the system.

In general, oxidation of the carbon portion of the organic materials provides a measure of this oxygen-depleting capability in any given aqueous media. Thus, analysis of the total organic carbon level in industrial and municipal wastewaters can provide an indication of the amount of organic pollutant present in the effluents.

Determination of organic carbon in pretreated sewage provides a measure of the amount of material still remaining that can deplete oxygen needed to sustain the life of delicate bacteria. Similarly, measurement of organic carbon serves as an indicator for all contaminants in seawater and in ship's bilge discharges. In addition, these measurements can be used in industrial processes to determine potential leakage of oil into product condensates or into cooling water and in the quality control of organic carbon levels in inorganic products.

In recent years, the growing need for continuous monitoring of total organic carbon in (1) industrial and municipal waste effluents, (2) studies relating to the performance of municipal waste treatment plants, (3) establishing the quality of potable and various process waters, (4) detection of oil spills, (5) controlling product losses into industrial waste effluents, (6) estimation of levels of organic compounds in boiler condensate, and (7) specification of organic contamination in inorganic products has necessitated the development of reliable instrumentation capable of analyzing total organic carbon from a wide variety of aqueous systems. Several methods, (U.S. Environmental Protection Agency, 1979; American Public Health Association, 1975) such as biochemical oxygen demand (BOD), chemical oxygen demand (COD), and total organic carbon (TOC), have been reported, which measure the concentration of organic materials in a wide variety of water systems. In addition, a procedure based on the ultraviolet measurement of dissolved organic pollutants in water has also been considered. Brief descriptions of these procedures are outlined in the following pages.

II. Biochemical Oxygen Demand

This procedure is commonly used for measuring waste loadings to treatment plants and in evaluating the efficiency of such treatment systems. The biochemical oxygen demand (BOD) method is also used to establish the relative oxygen requirements of treated effluents and polluted waters.

By definition, BOD represents the amount of oxygen utilized by a mixed population of microorganisms in the aerobic oxidation of organic compounds present in waste water systems at 20°C. The BOD procedure requires dilution of a known amount of wastewater sample with a prepared water in a BOD bottle. The prepared water for dilution contains phosphate buffer (pH 7.2), magnesium sulfate, calcium chloride, and ferric chloride. The mixture is further saturated with oxygen. The wastewater samples provide the organic carbon, and the dilution water supplies the dissolved oxygen. If sufficient microorganisms are not present in the test sample, seed microorganisms are added to oxidize the organic carbon. Depletion of dissolved oxygen due to the growth of bacterial population in the BOD bottle is directly related to the amount of degradable organic matters present in the test samples. The amount of oxygen depleted in the BOD bottle is obtained by a difference of initial and final measurement of dissolved oxygen. The oxygen is measured with either an oxygen-sensitive electrode or with the Winkler titration procedure over a five-day period at 20°C.

Owing to the fact that the growth rate of microorganisms is affected by environmental factors, such as substrate concentration, inorganic nutrient concentration, pH, temperature, toxic compounds, and types of microorganisms, the BOD procedure is not uniformly applicable to all types of organic wastes. In addition, the BOD method describes the rate of oxidation of organic matters under a given set of conditions, which may not match the physical, chemical, and biological stream conditions in real situations. The procedure requires five days for completion, and the results are often not reproducible (U.S. Environmental Protection Agency, 1979; American Public Health Association, 1975; Malcolm and Leenheer, 1973).

III. Chemical Oxygen Demand

The chemical oxygen demand (COD) procedure was developed to characterize the organic load in industrial wastewaters and pollution of natural waters. The method measures the amount of oxygen required for chemical oxidation of organic matter in the sample to carbon dioxide and water under acid conditions. A known amount of potassium dichromate in concentrated sulfuric acid is added to an Erlenmeyer flask containing known volumes of sample. The mixture is refluxed using a condenser for two hours. After cooling and diluting the mixture with organic carbon-free distilled water to a known volume, the remaining dichromate is titrated with standard ferrous ammonium sulfate. The amount of oxidizable organic material is proportional to the potassium dichromate used in the oxidation process, which, in turn, is proportional to the chemical oxygen demand. Semiautomated procedures (Pollution Control Laboratory, 1977; Adelman, 1966; Zuckerman and Molof, 1970) utilizing the principle of potassium dichromate oxidation of organic compounds in acid conditions have been reported.

The COD method determines total oxidizable organics but does not distinguish between biodegradable and nonbiodegradable organic materials. Furthermore, the procedure gives no indication of the rate at which organics would be oxidized in real sample streams. For these reasons, attempts to correlate COD with BOD values often require caution, except for a given waste stream, where an empirical relation between these parameters may be established.

IV. Ultraviolet Method

An alternative approach to BOD and COD methods, which appeared to have potential as a continuous monitor for organics, is based on the observations that most of the pollutants commonly found in water strongly absorb ultraviolet radiation. In fact, Dobbs *et al.* (1972) demonstrated a useful correlation between ultraviolet absorbance and TOC content for a variety of treated and untreated water samples. These samples ranged from municipal secondary sewage effluent to raw and processed river water. Sheppard (1976) studied the interferences of ultraviolet methods and claimed that the method is of limited value for analyzing organics in marine and estuarine systems. High concentration of inorganic UV absorbers, high concentration of organics that do not absorb in the ultraviolet, number and size of suspended particles, and turbidity of sample solution adversely affect the reliability of the ultraviolet method.

Middleton *et al.* (1962) reported a Mega Sampler capable of recovering organic contaminants from polluted water as a carbon chloroform extract (CCE). The water sample is passed through a sand filter, followed by a column of activated carbon. The absorbed materials are later removed by extraction, first with chloroform, then with alcohol. On concentration of extracts, a residue of organics present in the samples is obtained for ultraviolet measurement. However, in addition to being time-consuming, the ultraviolet procedures have not been adequately examined for their reliability and precision in actual field operations.

V. Total Organic Carbon

Because of the limitations of the BOD, COD, and UV absorbance methods and the present emphasis on automated water pollution control monitoring, the measurement of total organic carbon (TOC) is becoming an essential criterion for establishing water quality. In addition, the TOC procedure is proving to be more efficient for sample collection, storage, and analysis. Studies comparing TOC with BOD, COD, and UV analysis of the same samples are being made and correlations sought.

In general, for the determination of TOC in aqueous systems (Mitchell *et al.*, 1977), the samples are acidified, and dissolved CO_2 and CO_2 from carbonate and bicarbonate are removed by sparging. Organic carbon is then (1) oxidized by combustion or wet oxidation to CO_2 and detected using an infrared or a thermal conductivity detector or (2) oxidized, then reduced to methane and determined using a flame ionization detector (Van Hall *et al.*, 1965; Memzel and Vaccaro, 1964; Goulden and Brooksbank, 1975).

Reviews of the literature show that there is a growing interest in and need for continuous monitoring programs to measure TOC in a wide variety of water systems. Interest in the lower detection limits of the existing TOC analyzers also appears to be increasing.

In general, an automated TOC analyzing system must fulfill the following criteria to provide maximum benefit to the user:

1. fast analysis rate to maximize sample throughput,
2. longer instrument performance; e.g., calibration stability to minimize operator attention,
3. low sample and reagent consumption so that the system will be cost effective,
4. universal applicability so that special pretreatment of the sample is not required, and
5. simple maintenance and operation suitable for an unskilled operator.

The development of a continuous, on-line TOC analyzer system involves a few important considerations. For instance, first, a choice is made to use either low-temperature ultraviolet light-promoted chemical oxidation or high-temperature catalytic oxidation of organics present in sample matrices. Second, the design of the system is modified to eliminate interferences due to inorganic carbon, either by acidification and sparging or by measuring the amount of total carbon (TC) and inorganic carbon and taking the difference to obtain the total organic carbon level.

In the generation of TOC methods, questions always arise concerning the completeness of oxidation of organic molecules to carbon dioxide. Review of the literature shows that one can assume almost 100% oxidation for all types of samples, assuming that the samples under examination are potable water, condensate, cooling-tower water, municipal wastewater, and effluents from petroleum refineries, pulp, paper, and food processing plants. Although it is possible that certain samples may contain specific organic compounds that resist complete oxidation owing to their affect on the catalyst at high temperature, none has yet been reported in the literature, to the author's knowledge, which resist oxidation under ultraviolet-promoted chemical oxidation sequences.

VI. Ultraviolet-Promoted Chemical Oxidations

A marked trend toward wider use of ultraviolet-promoted chemical oxidation (digestion) techniques exists for several reasons. For instance, the technique assures (1) the complete oxidation of essentially any organic molecule, (2) that the digestion procedure will be simple and less complicated compared to high-temperature oxidation systems, (3) reduced maintenance and initial cost of the unit, and (4) easy oxidation of samples containing relatively high concentrations of salt.

To obtain complete oxidation of organics present in samples in a reasonable time, significant UV energy at wavelengths below 200 nm is required. The relative UV radiation from most mercury discharge lamps at the short wavelengths is small. Therefore, lamps of large power input are required to ensure sufficient energy at the short wavelengths. For instance, Goulden and Brooksbank needed a 1200-W lamp for complete oxidation of organic carbon in natural and lake water samples. Soier and Semenov (1971) studied the photochemical oxidation of 24 organic materials and reported that all of them were completely oxidized under the experimental conditions employed.

A necessary condition for a photochemical reaction to occur is that the molecule be excited by absorption of light quanta of sufficient energy (Grotthus-Draper law). Photochemical reactions usually occur in two stages: (1) the primary stage, which is directly due to the absorbed light quantum involving electronically excited states (this process is independent of temperature) and (2) the secondary stage, involving reactions of the various chemical species, molecules, ions, or radicals, that were produced by the primary photochemical reaction sequences.

Primary photochemical reactions can be initiated by physical factors, e.g., UV irradiation, ionizing radiation, heat, ultrasonic, or chemical factors, such as catalysis, direct reaction with oxygen, singlet oxygen, atomic oxygen, or ozone. Certain organic and inorganic compounds can induce and promote photochemical reactions by simply acting as either photoinitiators or photosensitizers. In photoinitiation processes, compounds are excited by absorbed light and subsequently decompose to free radicals. Photosensitizing activities are observed when molecules are excited by light and transfer their excitation energy to other molecules or to oxygen molecules forming singlet oxygen. In many cases, a chemical compound will, depending upon the condition of the experiment, behave either as a photoinitiator or a photosensitizer. It is well established that hydroperoxide groups form during photooxidation of many organic compounds. Since oxygen has a biradical nature, it combines rapidly with other free radicals, by addition, to form peroxy radicals, e.g.,

$$R\cdot + O_2 \rightarrow ROO\cdot.$$

The peroxy radical can abstract hydrogen from another organic molecule (R–H) to form hydroperoxide (ROOH) and a free radical to promote further chain reactions:

$$ROO \cdot + R-H \rightarrow ROOH + R \cdot.$$

The weak O–O bond of hydroperoxide is opened by UV irradiation to yield a variety of reactive free radicals:

$$ROOH \xrightarrow{h\nu} R \cdot + \cdot OOH,$$
$$ROOH \xrightarrow{h\nu} RO \cdot + \cdot OH.$$

These free radicals take part in radical induced decomposition of hydroperoxides and disproportionation sequences, causing degradation of organic molecules:

$$ROOH + RO \cdot \rightarrow ROO \cdot + ROH,$$
$$ROOH + HO \cdot \rightarrow ROO \cdot + H_2O.$$

Molecular oxygen exists in the ground state as triplet state ($^3O_2$) and as singlet oxygen ($^1O_2$) in excited states. UV irradiation of oxygen molecules generates ($^1O_2$), which, in turn, reacts with organic molecules to form rapidly degradable intermediates, such as hydroperoxides and endoperoxides. In addition, most water systems are likely to contain certain metals, either as salt or as metal-organic complexes. Under UV irradiation, these metals react with molecular oxygen to generate reactive free radicals and ions, which cause accelerated degradation of organic molecules:

$$M^{n+} + O_2 \rightarrow M^{(n+1)+} + \dot{O}_2^-,$$
$$\dot{O}_2^- + H^+ \rightarrow H\dot{O}_2,$$
$$\dot{O}_2^- + RH \rightarrow RO \cdot + HO^-.$$

Metal ions also decompose hydroperoxides to form oxidative free radicals:

$$M^{n+} + ROOH \rightarrow M^{(n+1)+} + RO \cdot + HO^-,$$
$$M^{(n+1)+} + ROOH \rightarrow M^{n+} + ROO \cdot + H^+.$$

Baldwin and McAtee (1974) used silver-catalyzed peroxydisulfate at room temperature to oxidize a variety of natural water samples. Peroxydisulfate in an aqueous solution behaves as a strong oxidizing agent (House, 1962). The oxidizing power of peroxydisulfate is greatly enhanced when it is allowed to decompose either in the presence of metal ions, such as silver and copper, or UV irradiation. Under these conditions, peroxydisulfate first decomposes to generate sulfate-free radicals (SO_4^-), which in turn, react with water to produce hydroxyl radicals, e.g.,

$$S_2O_8^{2-} \rightarrow 2S\dot{O}_4^{2-} \qquad \text{(uncatalyzed)},$$
$$S_2O_8^{2-} + Ag^+ \rightarrow Ag^{2+} + S\dot{O}_4^{2-} + SO_4^{2-} \qquad \text{(silver ion catalyzed)},$$
$$S_2O_8^{2-} + Cu^{3+} \rightarrow Cu^{3+} + S\dot{O}_4^{2-} + SO_4^{2-} \qquad \text{(copper ion catalyzed)},$$
$$S\dot{O}_4^{2-} + H_2O \rightarrow HSO_4^- + \cdot OH.$$

Depending on the experimental conditions employed for UV-catalyzed digestion of samples, the generated ·OH radicals either initiate chain reactions for the decomposition of organic molecules or scavange the sulfate-free radicals, e.g.,

$$S\dot{O}_4^- + HO\cdot \rightarrow SO_4^{2-} + \frac{1}{2}O_2.$$

In acidic conditions, peroxydisulfate first decomposes to give peroxymonosulfuric acid, which hydrolyzes further to hydrogen peroxide. Decomposition of hydrogen peroxide, either with UV irradiation or catalysis by metal ions, provides a source of free ·OH radicals.

VII. Automated Total Organic Carbon Analysis

Several procedures that employ either UV irradiation or silver-catalyzed peroxydisulfate oxidation of organic compounds have been considered for automation. For instance, Erhardt (1969) reported an automated procedure for the analysis of seawater samples using irradiation of samples with ultraviolet light in the presence of peroxydisulfate. The carbon dioxide produced is absorbed in alkali solution. The measurement of conductivity of the solution gives the concentration of dissolved carbon dioxide. Goulden and Brooksbank (1975) described an automated system using either UV irradiation or silver-catalyzed peroxydisulfate to affect oxidation of samples. An infrared analyzer was used to measure the generated carbon dioxide.

The Technicon TOC Monitor IV® System, which is designed for continuous, on-line monitoring of TOC, DOC, and TC from a wide variety of aqueous sample media, uses a low-pressure, 14-W mercury lamp to provide ultraviolet irradiation of samples. The UV lamp contains a quartz envelop, which allows unrestricted passage of the 185-nm Hg/ozone spectral lines. The instrument operates on the principle of continuous-flow analysis, where air-segmented streams of sample and reagent are mixed together and allowed to react under carefully controlled conditions. For determination of TOC and DOC, the sample is acidified and sparged to eliminate inorganic carbon. A portion of the inorganic carbon-free sample is resampled and mixed with sulfuric acid and potassium persulfate. The mixture is irradiated with ultraviolet light in a UV Digestor (the unit contains a quartz coil, wrapped around the UV lamp). Carbon dioxide, produced by oxidation, is separated from the sample matrix by passage through a dialyzer containing a silicone rubber gas-permeable membrane. The recipient stream contains phenolphthalein, which is dissolved in carbonate–bicarbonate buffer. The change in color as the pH changes by the absorption of CO_2 is measured with a colorimeter.

The chemical reaction sequences employed in the Technicon Monitor IV System for the measurement of total organic carbon in water, wastewater, and

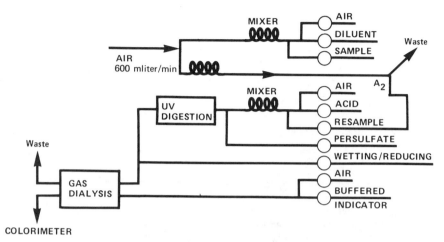

Fig. 1. Flow diagram showing the chemical reaction sequences in monitor IV systems.

process effluents can be shown by a simplified flow diagram (Fig. 1). For the measurement of total carbon, the sparging step is omitted and the samples are analyzed directly. The functional characteristics (DiLiddo, 1977) of the TOC Monitor IV System are summarized in Table I. Selected groups of samples were analyzed (DiLiddo, 1977) and the results compared with the total carbon combustion method. Comparison of the results indicated that the Monitor IV provides (Table II) accurate measurement of total organic carbon.

During sparging of samples to remove inorganic carbon, certain organic compounds may be lost, especially when they are both volatile and extremely insoluble in an aqueous medium. This loss can be readily determined by comparing the

TABLE I

FUNCTIONAL CHARACTERISTICS OF
TECHNICON TOC MONITOR IV® SYSTEM

Range of analysis:	2–100 or 10–500 mg C/liter
Baseline drift:	0.002 AU/°C
Span drift:	Less than 1.0% of full scale/24 hr
Noise:	Less than 1.0% of full scale
Linearity:	Maximum deviation 1.0% of full scale
Detection limit:	2 mg C/liter or 10 mg C/liter
Response time:	Lag time: 7 min for TOC
	5 min for TC
Maintenance requirements:	(a) Pump tube changes: monthly
	(b) Reagent replenishment: weekly or monthly
	(c) Dialysis membrane changes: monthly

TABLE II

COMPARISON OF ANALYTICAL DATA OBTAINED BY COMBUSTION AND MONITOR IV METHODS

Sample type	Combustion method (mg C/liter)	Monitor IV method (mg C/liter)	Recovery (%)
Process influent (paper industry)	450	450	100
Industrial process effluent	52	51.5	99
Industrial effluent (petrochemical)	10	9.6	96
Industrial effluent (petrochemical)	115	110	96
Sewage effluent	66	67	101
Sewage influent	48	49	102

readings obtained with and without sparging on an unacidified sample. Acidified samples will give a lower reading by the amount of inorganic carbon present. In general, it is rare in wastewater analysis that highly volatile and extremely insoluble materials are present, but the probability is greater for other types of samples.

To best determine the effectiveness on recovery using the Technicon Monitor IV System, recovery data are listed in Table III using selected pure organic compounds both in the presence and absence of sodium chloride. A known standard potassium hydrogen phosphate (KHP) solution was used for instrument calibration prior to analysis of organic compounds. Acetonitrile and dichloroethane are volatile compounds, and a small percentage is lost during the sparging step. In addition, by changing the digestion reagent from acid persulfate to neutral phosphate buffer, the effects of sodium chloride, when present in amounts up to 25% as an interferent, are eliminated completely.

TABLE III

PERCENT RECOVERY OF SELECTED ORGANIC COMPOUNDS BOTH IN THE PRESENCE AND ABSENCE OF 25% SODIUM CHLORIDE

Compound	Recovery without sodium chloride	Recovery with sodium chloride
Pyridine	100	100
1-Butanol	99	99
Acetic acid	102	102
Proline	100	100
Acetonitrile	82	82
Glutamic acid	100	[a]
Dichloroethane	93	[a]
p-Nitrophenol	100	100

[a] Sample not analyzed.

VIII. Description of Technicon Total Organic Carbon
Monitor IV System

The basic components of the Technicon Monitor IV® System (Fig. 2) are (1) the overflow sampler, (2) the solenoid valve, (3) the proportioning pump, (4) the UV digestor, (5) the manifold assembly, (6) the detector, (7) the recorder, and (8) the sparging system. Owing to a complete modular design of the instrument (Fig. 3), the maintenance and replacement of components are easily accomplished.

The design and functional operation of the above components are briefly described below.

1. Overflow sampler. This unit allows an aliquot of liquid sample to be introduced into the main manifold by the proportioning pump, while excess sample flows to waste.

2. Solenoid valve. This module, energized by a 24-hr timer, is capable of introducing either liquid samples or reference solution into the analytical manifold.

3. Proportioning pump. This unit moves samples and reagents through the entire system. Basically, the module consists of two parallel stainless steel roller

Fig. 2.   Technicon TOC monitor system.

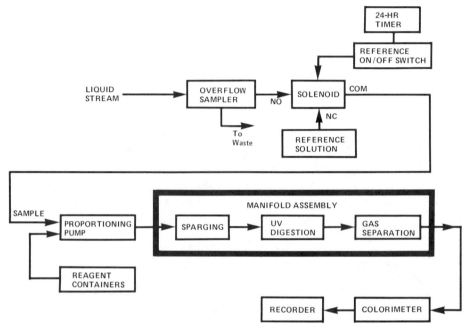

Fig. 3. Block diagram showing the interchangeable modular designs of monitor IV systems.

chains connected by five equally spaced stainless steel rollers, a constant-speed motor, which drives the chain assembly, and a spring-supported platen. Resilient pump tubes, held tight by two end blocks, are placed between the platen and rollers. As the rollers move over the pump tubes, pressure occludes the tubes at points of contact. This action pushes the fluids forward through the tubes and draws more fluid into the tubes from behind.

4. UV digestor. This unit contains an ultraviolet lamp, a low-pressure, 14-W Hg lamp that operates at relatively low temperatures. Both the ultraviolet lamp and the coil carrying the sample for irradiation with ultraviolet light are made of quartz. Owing to the unique built-in design of the UV Digestor, complete oxidation of organic molecules to carbon dioxide is virtually assured.

5. Manifold assembly. The manifold assembly contains necessary glassware, tubing, heating baths, dialyzers, and fittings. All manifold components (except special auxiliary distilling and pumping units) are housed in a single modular unit, with removable, see-through cover. The Monitor IV unit is designed to facilitate the change of manifold assemblies for different analyses.

6. Detectors. The Monitor IV System incorporates either a colorimeter or ion selective electrode (ISE), depending on the parameters analyzed. The colorimeter is provided with sample and reference channels with a common excitation

source and discrete phototube detectors for operation in the range of 340–880 nm with appropriate filters. It is capable of accommodating flow cells of either 15- or 50-mm pathlength.

7. Recorder. The Monitor IV unit is capable of providing both a direct analog data output and a stripchart printout for permanent records of analytical values.

8. Sparging system. The unit consists of an impingement pump, sparging coil, and an additional gas–liquid separator. This module removes inorganic carbon from the sample solution. Samples are acidified with 1 N H_2SO_4 and sparged with purified ambient air. The air for sparging is delivered by an air pump and purified from organic contaminants by passing through a filter containing activated charcoal. The sparging coil can be bypassed if measurements of total carbon (inorganic and organic) are desired.

In addition, the Technicon TOC Monitor IV options are available for the following additional functions:

(a) *Dual sample analysis.* By addition of a timer and a programmable valve, two separate streams can be analyzed alternately with the capability to automatically standardize (autocorrect) the system at predetermined intervals. An automatic correction (autocorrection) feature is provided for monitoring applications exhibiting drift, due to inherent system changes, such as reagent degradation, pump tube delivery changes, and electronic long-term drifts. Depending upon the expected nature of drift, the automatic correction unit may be set to correct for either baseline drift or sensitivity drift.

(b) *High–low alarms.* This module alerts the operator to any unusual change in the concentrations of the parameter being monitored.

(c) *Continuous water clarifier.* This unit provides prefiltered samples to particulates larger than 0.5 μm from sample streams.

IX. Conclusions

Because of a growing desire to improve our water resources by monitoring the influx of industrial, agricultural, and municipal pollutants, the application of TOC measurements have widened significantly. The literature (Jones, 1972; Chandler *et al.*, 1976) shows that some valid relationships exist between TOC and BOD or COD measurements. In order to replace or complement the existing BOD and COD methods with the rapid TOC procedure, which is applicable to continuous on-line monitoring of samples, it is necessary to study the way in which TOC measurements relate to BOD and COD for different types of aqueous samples. Such studies necessitate the performance of TOC analysis on a wide variety of aqueous systems. In addition, the application of TOC data are expected to expand with the various industrial processings and other related industries.

Technicon TOC Monitor IV is a complete system, one that provides rapid on-line measurement for total organic carbon (TOC), dissolved organic carbon (DOC), and total carbon (TC). Fully automated operation and long-term instrument stability reduce operator error and allow operation by unskilled personnel. The Technicon Monitor IV System has flexibility, an advanced design, and the capability of providing precise, accurate, and reliable TOC data.

References

Adelman, M. H. (1966). Simplified automated COD determination: Advanced procedures. *Autom. Anal. Chem., Technicon Symp., 1965* pp. 552–556.

American Public Health Association (1975). "Standard Methods for the Examination of Water and Waste Water." American Water Works Association and Water Pollution Control Federation, Washington, D.C.

Baldwin, J. M., and McAtee, R. E. (1974). Determination of organic carbon in water with a silver-catalyzed peroxydisulfate wet chemical oxidation method. *Microchem. J.* **19,** 179–190.

Chandler, R. L., O'Shaughnessy, J. C., and Blank, F. C. (1976). Pollution monitoring with total organic carbon analysis. *J.—Water Pollution Control Fed.* **18**(2), 2791–2803.

DiLiddo, J. (1977). On-line monitoring for organic carbon detection. *Adv. Autom. Anal. Technicon Int. Congr., 1976* Vol. 2, pp. 22–26.

Dobbs, R. A., Wise, R. H., and Dean, R. B. (1972). The use of ultraviolet absorbance for monitoring of total organic content of water and wastewater. *Water Res.* **6,** 1173–1180.

Erhardt, M. (1969). A new method of the automatic measurement of dissolved organic carbon in sea water. *Deep-Sea Res.* **16,** 393–397.

Goulden, P. D., and Brooksbank, P. (1975). Automated determination of dissolved organic carbon in lake water. *Anal. Chem.* **47,** 1943–1946.

House, D. A. (1962). Kinetics and mechanisms of oxidations by peroxydisulfate. *Chem. Rev.* **62,** 185–203.

Jones, R. H. (1972). TOC: How valid is it? *Water Waste Eng.* **9,** 32–33.

Malcolm, R. L., and Leenheer, J. A. (1973). The usefulness of organic carbon parameters in water quality investigations. *Proc. Annu. Tech. Meet.—Inst. Environ. Sci.* **19.**

Memzel, D. W., and Vaccaro, R. F. (1964). The measurement of dissolved organic and particulate carbon in sea water. *Limnol. Oceanogr.* **9,** 138–142.

Middleton, F. M., Pettit, H. H., and Rosen, A. A. (1962). The mega sampler for extensive investigation of organic pollutants in water. *Proc. Ind. Waste Conf.* **17,** 454–460.

Mitchell, D. G., Aldous, K. M., and Canelli, E. (1977). Determination of organic carbon by thermal volatization—plasma emission spectrometry. *Anal. Chem.* **49,** 1235–1238.

Pollution Control Laboratory (1978). "Methods Manual for Chemical Analysis of Water and Wastes." Revised Method NAQUADAT No. 08304L. Water Analysis Section, Alberta Environment, Edmonton, Alberta.

Sheppard, C. R. C. (1976). Problems with the use of ultraviolet absorption for measuring carbon compounds in a river system. *Water Res.* **11,** 979–982.

Soier, V. G., and Semenov, A. D. (1971). Photochemical method for determining organic carbon. *Gidrokhim. Mater.* **56,** 111–120; *Chem. Abstr.* **25,** 121236P.

U.S. Environmental Protection Agency (1979). "EPA: Methods for Chemical Analysis of Water and

Waste Waters, 1979," Method Nos. 405.1, 410.1-.2, .3, .4, and 415.1. Environmental Monitoring and Support Laboratory, Office of Research and Development, USEPA Cincinnati, Ohio.

Van Hall, C. E., Safranko, J., and Stenger, V. A. (1965). Rapid combustion method for the determination of organic substances in aqueous solution. *Anal. Chem.* **35,** 315–319.

Zuckerman, H. M., and Molof, A. H. (1970). Wastewater renovation for reuse studied by the use of extremely low level automated chemical oxygen demand. *Adv. Autom. Anal., Technicon Int. Congr., 1969* pp. 121–124.

Index

A

Albumin analysis, in clinical chemistry laboratory, 212–213

Alcohols, in salt brines, 7

Aluminum in nuclear power plant coolant, analysis by flameless atomic absorption, 87–88

Ammonia in water, flow-injection method, 55–58

Analytical process in the clinical laboratory, 192

Anions, by ion chromatography, 17–18

Applications, for liquid chromatographic analyses, 185

B

Basic oxygen sampling and analyzing systems, 297

Blast furnace gas, schematic of analyzer system, 292

Blast furnace gas sampling system, 291

Boric acid in nuclear power plant coolant, 70
analysis by flameless atomic absorption, 83–85

n-Butane from coal conversion chromatographic analyses, 109

2-Butane from coal conversion chromatographic analysis, 109

C

Calcium in milk, flow-injection method, 55

Calcium in nuclear power plant coolant, analysis by flameless atomic absorption, 87–88

Carbon dioxide
from blast furnace gas, by infrared analyzers, 284–285
from coal conversion, chromatographic analysis, 109

Carbon monoxide
from blast furnace gas, infrared analysis, 284–286
from coal conversion, chromatographic analysis, 109

Cations
by flameless atomic absorption, 69–75
by ion chromatography, 30–33

Centrifugal analyzers, in clinical chemistry laboratory, 227

Classification of automated analysis, in clinical chemistry laboratory, 203

Clinical chemistry laboratory
automated high-performance liquid chromatography, 119–187, 210–211
automated radioimmunoassay, 207–210, 224–226
blood gas analyzers, 235
centrifugal analyzers, 226–228
classification of automated analysis, 202–204
discrete analysis in open tubes, 213–216
discrete analyzer with optical scanning, 220–222
flow-injection analysis, 39–67, 211–213
high-volume multichannel analyzers, 216–220
multichannel in situ analyzer, 234
prepackaged single-test reagents, 228–231
segmented flow analysis, 39–67, 204–210